SpringerBriefs in Computer Science

W0193219

Series Editors

Stan Zdonik
Peng Ning
Shashi Shekhar
Jonathan Katz
Xindong Wu
Lakhmi C. Jain
David Padua
Xuemin Shen
Borko Furht
V.S. Subrahmanian
Martial Hebert
Katsushi Ikeuchi
Bruno Siciliano

For further volumes:
http://www.springer.com/series/10028

Ning Zhang • Jon W. Mark

Security-aware Cooperation in Cognitive Radio Networks

 Springer

Ning Zhang
Department of Electrical
   and Computer Engineering
University of Waterloo
Waterloo, ON, Canada

Jon W. Mark
Department of Electrical
   and Computer Engineering
University of Waterloo
Waterloo, ON, Canada

ISSN 2191-5768   ISSN 2191-5776 (electronic)
ISBN 978-1-4939-0412-9   ISBN 978-1-4939-0413-6 (eBook)
DOI 10.1007/978-1-4939-0413-6
Springer New York Heidelberg Dordrecht London

Library of Congress Control Number: 2013958216

Printed on acid-free paper

Springer is part of Springer Science+Business Media (www.springer.com)

Preface

Cognitive radio networks (CRNs) are envisaged to solve the problem of spectrum scarcity, caused by the current spectrum allocation policy in which only licensed users have channel access rights. In CRNs, unlicensed users are allowed to opportunistically use the idle spectrum bands owned by licensed users to meet the ever-increasing demand on spectrum and increase spectrum efficiency. In order to access the spectrum bands without creating interference to the licensed users, unlicensed users need to conduct spectrum sensing. However, spectrum sensing might be inaccurate due to multipath fading, shadowing, and primary receiver uncertainty. To address this problem, two types of cooperation have been proposed in the literature: cooperative spectrum sensing and cooperative cognitive radio networking (CCRN). For the former, the cooperation is performed among unlicensed users; while for the latter, the cooperation is carried out between unlicensed users and licensed users. As an emerging paradigm, CCRN can achieve mutual benefits for both participants, which will be the focus of this book. Whereas cooperation can also incur security issues, e.g., malicious users might participate in the cooperation to corrupt or disrupt the communication of legitimate users. Those security issues are of great importance and need to be addressed before the widespread deployment of cooperation in CRNs.

In this book, we study cooperative networking in CRNs, where unlicensed users and licensed users cooperate with each other to obtain mutual benefits, taking security aspects into consideration. In Chap. 1, we first give a brief introduction to CRNs, including fundamentals of cognitive radio, spectrum sensing, and cooperation in CRNs. In Chap. 2, a literature survey on cooperative networking in CRNs is provided, followed by a discussion on the security aspects, which also motivate the subsequent works in Chaps. 3 and 4. Specifically, a trust-aware cooperation scheme for CRNs to improve throughput or energy efficiency of licensed users and offer transmission opportunities to unlicensed users, considering the trustworthiness of unlicensed users, is presented in Chap. 3; and a cooperation scheme to enhance secure communications of licensed users is presented in Chap. 4. Numerical results are provided also to evaluate the proposed schemes. Finally, the concluding remarks are given in Chap. 5.

The authors would like to thank Prof. Xuemin (Sherman) Shen, Ning Lu, and Nan Cheng of the Broadband Communications Research Group (BBCR) at the University of Waterloo, and Prof. Rongxing Lu of Nanyang Technological University, for their contributions in the presented research works. Special thanks are also due to the staff at Springer Science+Business Media: Courtney Clark and Jennifer Malat, for their help throughout the publication preparation process.

Thanks also to the first author's parents, Jianguang Zhang and Jianhua Duan, and his wife, Xue Qin, for their love and support.

Waterloo, ON, Canada   Ning Zhang
Waterloo, ON, Canada   Jon W. Mark

Contents

Acronyms

AF	Amplify-and-forward
BS	Base station
C-B	Cluster-beamforming cooperation scheme
CBSE	Cluster-beamforming scheme for single eavesdropper
CBME	Cluster-beamforming scheme for multiple eavesdroppers
CCRN	Cooperative cognitive radio networking
CR	Cognitive radio
CRNs	Cognitive radio networks
CSI	Channel state information
DF	Decode-and-forward
DSA	Dynamic spectrum access
DSTC	Distributed space-time coding
E-CSI	Eavesdropper's channel state information
FDMA	Frequency-division multiple access
MRC	Maximal ratio combining
NE	Nash equilibrium
PDF	Probability density function
PHY	Physical layer
PU	Primary user
QoS	Quality of service
R-J	Relay-jammer cooperation scheme
SU	Secondary user
SNR	Signal-to-noise ratio
TDMA	Time division multiple access

Chapter 1
Introduction

Abstract Cognitive radio is a promising solution to the spectrum scarcity versus underutilization dilemma, which enables unlicensed users to opportunistically use the unused spectrum owned by licensed users to increase the spectrum utilization. To avoid interfering with the operation of licensed users, unlicensed users perform spectrum sensing before transmission to detect the available channels. However, the outcome of spectrum sensing may be inaccurate due to fading or shadowing. To overcome this problem, cooperation has been leveraged in cognitive radio networks (CRNs). In this chapter, we first give a brief introduction to CRNs, including fundamentals of cognitive radio, spectrum sensing, and cooperation in CRNs.

1.1 Cognitive Radio

In recent decades, it has been witnessed a rapid growth in wireless communications, which have been almost applied to every aspects of personal. Emerging wireless devices and applications further accelerate the development of wireless systems. Such an exponential growth of wireless communication also imposes huge demands on radio spectrum. As a natural resource, radio spectrum is scarce and limited. Nowadays, the spectrum is managed by government agencies such as the Federal Communications Commission (FCC), and assigned to licensed users on a long term basis to avoid interference among wireless systems. Although this static allocation approach worked well in the past, it cannot serve the ever increasing demand for wireless communication well because of the problem of spectrum scarcity. Recent studies reveal that the allocated spectrum are underutilized. As shown in Fig. 1.1, some parts of spectrum remain largely underutilized, some parts are sparingly utilized, while the remaining parts of the spectrum are heavily occupied [1]. It is recognized that this kind of static allocation policy has resulted in poor spectrum utilization, and created a severe shortage of spectrum for unlicensed users. Furthermore, spectrum underutilization by licensed users exacerbates spectrum scarcity. The main reason of spectrum underutilization is that licensed users

N. Zhang and J.W. Mark, *Security-aware Cooperation in Cognitive Radio Networks*,
SpringerBriefs in Computer Science, DOI 10.1007/978-1-4939-0413-6__1,
© The Author(s) 2014

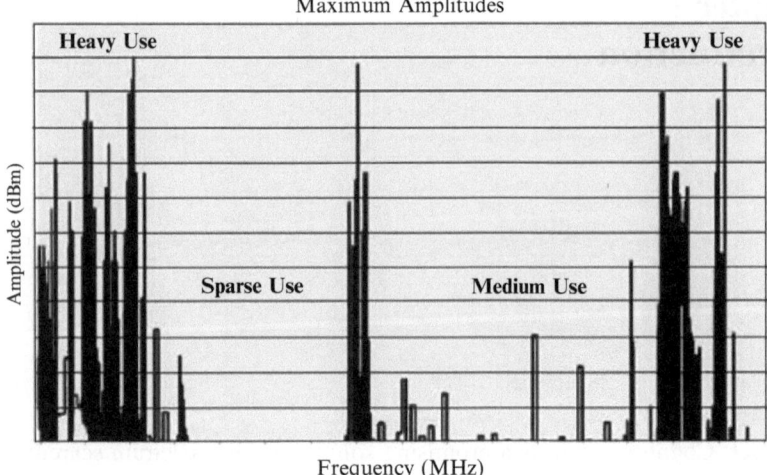

Fig. 1.1 Spectrum utilization [1]

typically do not fully utilize their allocated bandwidths for most of the time, while unlicensed users are being starved for spectrum availability. To deal with this dilemma, cognitive radio is a paradigm created in an attempt to enhance spectrum utilization, by allowing unlicensed users to coexist with licensed users and make use of the spectrum holes [1–3]. The spectrum holes are defined as the spectrum bands owned by licensed users, which are unused at a particular time and specific geographic location.

Cognitive radio (CR) is defined as a radio that can sense the surrounding wireless environments where it operates and adjust the transmission parameters accordingly. To be more precise, FCC gives the definition as follows: "*Cognitive radio: A radio or system that senses its operational electromagnetic environment and can dynamically and autonomously adjust its radio operating parameters to modify system operation, such as maximize throughput, mitigate interference, facilitate interoperability, access secondary markets.*" [4].

Two main characteristics that distinguish CR from the traditional wireless radio are cognitive capability and reconfigurability. The former represents the awareness of CR with respect to the transmitted waveform, RF spectrum, communication network, geography, locally available services, user need, security policy and so on, while the latter corresponds to capability of adaption to the obtained information about the wireless environments [3].

To better understand cognitive radio, software-defined radio (SDR) is briefly reviewed. Introduced in 1991, SDR is defined as a radio platform, whereby components implemented in hardware (modulation/demodulation, compression, filtering, error correction coding, etc.) can be implemented by means of software instead. Thanks to the rapid development of microelectronic, wireless devices become more and more powerful and capable, which in return facilitate the evolution of SDR.

With SDR, the functionalities or the operational parameters of the devices can be programmable, which enables reconfiguration of the radio by just changing the codes. With the feature of reconfigurability, many new applications emerge. For instance, SDR can be used to build radios to support multiple interfaces by reconfiguring them in software. In addition, the base transceiver station (BTS) or other devices, implemented using SDR, can be upgraded simply without too much expense, e.g., from Global System for Mobile Communication (GSM) to General packet radio service (GPRS).

Although SDR can provide the reconfigurable capability, it can only be performed on demand. In other words, SDR cannot carry out reconfiguration itself. Different from SDR, cognitive radio can achieve self-configuration through learning from the environment where it operates due to the fundamental characteristics of cognitive capability and reconfigurability. With cognitive capability, cognitive radio can acquire the information from the environment and adapt without being programmed a priori. With the reconfigurability, cognitive radio can be dynamically programmed accordingly, which is implemented on the platform of SDR. In other words, cognitive radio can automatically sense or detect the radio environment, and change its transmission or reception parameters accordingly such that the resource can be utilized in an efficient way.

1.1.1 Functions

With CR technology, unlicensed users can coexist with licensed users to exploit the spectrum holes. However, there is a stringent requirement on the unlicensed users. That is, the operation of unlicensed users must not interfere with the transmissions of licensed users. Moreover, considering the diverse quality of service (QoS) requirements of unlicensed users, CR faces many challenges. To overcome those challenges, the following functions of CR are identified: spectrum sensing, spectrum decision, spectrum sharing, and spectrum mobility [5].

Spectrum sensing is a very important function, which should be performed to acquire information from the surrounding environment, such as presence of the licensed users and channel availabilities, before transmission [6, 7]. It is necessary for CR to adapt its operational parameters according to the status of the environment. The objectives of spectrum sensing can be classified as follows: (1) the operation of unlicensed users should avoid harmful interference to licensed users by either switching to an available band or limiting its interference to licensed users at an acceptable level, and (2) unlicensed users should efficiently and reliably identify the spectrum holes to meet their QoS requirements. Therefore, spectrum sensing is crucial for both licensed and unlicensed users.

After available channels are detected, spectrum decision is performed to select suitable channels according to the QoS requirement of unlicensed users. The decision is made based on the results of spectrum sensing and the internal policy of the users (e.g., to maximize throughput, reliability, or have the longest transmission time, and so on). Subsequently, the best channel to access is selected among available channels.

When multiple unlicensed users exist, spectrum sharing is necessary, especially for distributed CRNs. Spectrum sharing refers to the process of sharing the common available channels among multiple unlicensed users. The objective is to utilize the available channels in an efficient and fair way by coordinating the users [8–10].

Since an unlicensed user has to vacate the current channel once the presence of licensed user is detected, the unlicensed user has to find another available channel to access in order to maintain the ongoing transmission. This process is referred to as spectrum mobility. The goal is to meet the QoS requirement of unlicensed users by means of choosing the channel to move or sense.

1.1.2 Applications

Since CR is capable of autonomously adapting its operational parameters (e.g., transceiver parameters) to work in a more efficient way, based on the information acquired from the environment by active monitoring, a large number of promising applications can be facilitated, ranging from military to commercial market. Two key applications, dynamic spectrum access [11–15] and interoperability [16, 17], are identified in the academia and industry. Dynamic spectrum access refers to the scenario where unlicensed users sense the available channels which are not occupied by licensed users, and then access those channels for transmission. Interoperability means that radios can connect different systems operating on different protocols or standards so that they can communicate with each other.

1.1.2.1 Dynamic Spectrum Access

Dynamic spectrum access (DSA), as shown in Fig. 1.2, is the main application, which has received great attention from both academia and industry. The objective of DSA is to efficiently utilize the spectrum to solve the problem of spectrum secrecy, which is the result of the ever increasing mobile devices and the current static spectrum allocation policy. DSA allows unlicensed users to opportunistically utilize the licensed spectrum bands when they are unoccupied. In order to avoid harmful interference to the legacy system, unlicensed users have to carry out spectrum sensing to detect the spectrum holes. Once the available channels are detected, unlicensed users can access for their transmission. During transmission, spectrum sensing has to be continuously performed to sense the activities of licensed users. When the presence of a licensed user is detected, the unlicensed user vacates the current channel and chooses other channels to sense for transmission opportunities.

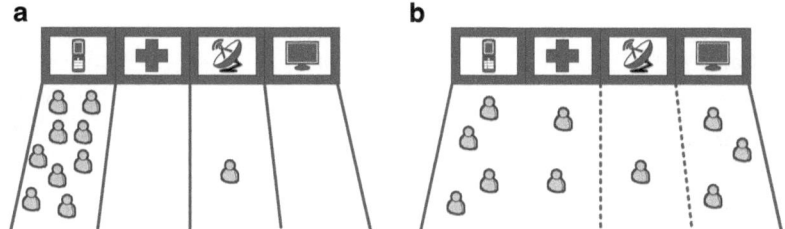

Fig. 1.2 Dynamic spectrum access. (**a**) Wireless systems without CR, (**b**) wireless systems with CR

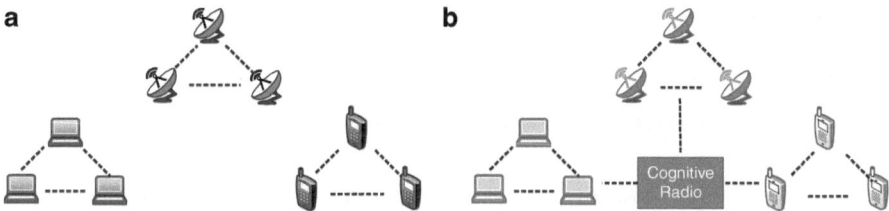

Fig. 1.3 Interoperability enabled by cognitive radio. (**a**) Wireless systems without CR, (**b**) wireless systems with CR

1.1.2.2 Interoperability

As the second important application, interoperability has a huge potential impact on the current communication architecture, which is expected to further affect the personal life of human. Nowadays, we are surrounded by different types of communication systems, e.g., mobile networks, sensor networks, wireless local area network, TV broadcast network, and so on. Those systems are independent and autonomous systems, with different standards, spectrum bands, services, etc. With the technology of cognitive radio, the device can reconfigure itself to communicate with incompatible radios. Specifically, CR first scans the surrounding environment to detect what waveforms or networks are present. Then, it can either reconfigure itself to communicate with the selected network, or it can act as a gateway to connect different systems for communications. By doing so, diverse wireless systems, as shown in Fig. 1.3, can be connected and communicate with each other.

1.1.3 Network Architecture

With the assistance of cognitive radio technology, unlicensed users can coexist with licensed users and utilize the temporarily unused spectrum bands owned by licensed users. Therefore, CR network architecture is comprised of two components: the primary network and the secondary network, as shown in Fig. 1.4. Both networks

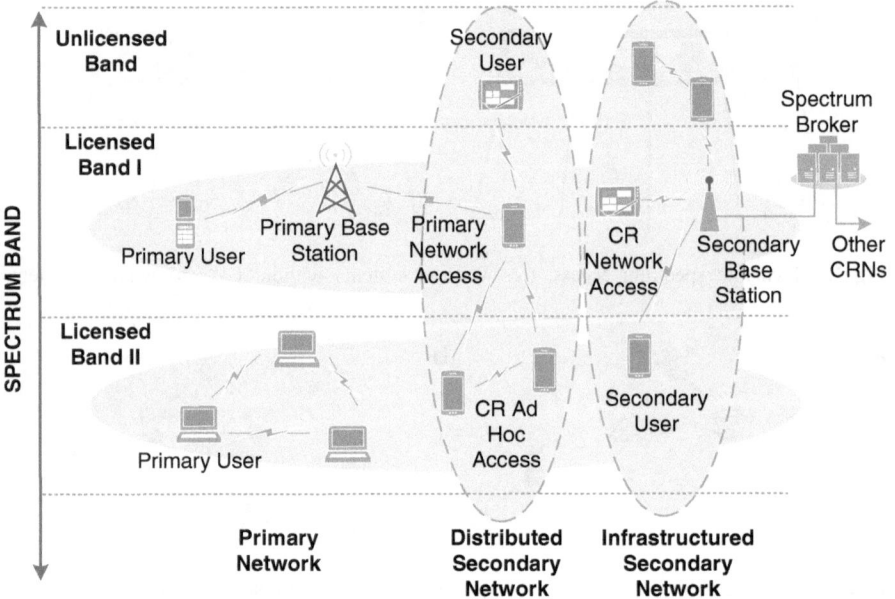

Fig. 1.4 Cognitive radio network architecture [5]

can be deployed in either a centralized or ad hoc mode, where communications are coordinated by central nodes such as base stations or communications are carried out in a peer-to-peer fashion, respectively. In cognitive radio networking, unlicensed users are referred to as secondary users (SUs), while licensed users are coined to as primary users (PUs).

1.1.3.1 Primary Network

The primary network corresponds to an existing network which holds a license for operation in certain spectrum bands. This network has the exclusive privilege to access the assigned spectrum bands. If the primary network has an infrastructure, PUs can be coordinated to access the network through the primary base station. In addition, the primary network might be deployed in ad hoc mode, where PUs communicate with each other without any infrastructure. The PUs' transmissions occurring in the primary network should be protected from being interfered by secondary networks. Generally speaking, PUs and primary base stations are typically not equipped with CR functions. Therefore, it is the responsibility of SUs to sense the channel before transmission and vacate the occupied channel when PUs re-appear.

1.1.3.2 Secondary Network

The secondary network, composed of a set of SUs, does not have the license to operate in any licensed spectrum bands. The secondary network can also be classified into two types: infrastructure-based and ad hoc [18]. An infrastructure-based secondary network has a central controller, e.g., a secondary base station or an access point. Opportunistic spectrum access by SUs is usually coordinated by the central controller. Whereas in an ad hoc secondary network, SUs can communicate with each other via multi-hop wireless links on either the licensed or the unlicensed spectrum bands. Both SUs and secondary base stations are equipped with CR technology. When different secondary networks share one common spectrum band, a spectrum broker is needed to coordinate them.

1.1.4 Operational Modes

The operational modes include interleave, underlay, and overlay [19], as detailed below.

1.1.4.1 Interleave

For interleave mode, SUs has to carry out spectrum sensing to detect the unused channels before transmission. Only when there is no active PU in the channel, can an SU access for transmission. During transmission, spectrum sensing has to be performed to check whether or not the PU has returned. Once the presence of a PU is detected, the SU should vacate the current operating channel and sense to find another idle channel for transmission. This mode has the disadvantage of being sensitive to detection errors and the PUs' activities.

1.1.4.2 Underlay

The underlay mode allows the SU to transmit data simultaneously with the PU under the condition that the interference caused by the SU at the primary receiver should be below a predefined threshold. It is different from the interleave mode, where the interference is completely avoided. However, this mode relies on an accurate estimate of the interfering channel, which in reality is hard to obtain. Moreover, the SU has poor performance because of the following reasons: (1) due to the interference limit, the SU has a power constraint which affects its performance; (2) the SU's transmission also suffers from interference by the PU's transmission.

1.1.4.3 Overlay

The overlay mode leverages cooperative networking to exploit transmission opportunities, whereby the SU cooperates with the PU to enhance the PU's transmission performance in terms of throughput, reliability, and so on, and in return the PU grants a interval of time to the SU for its own transmission. This mode is actually based on mutual benefits, i.e., both participants get benefits from cooperation. In the literature, this mode is also called cooperative cognitive radio networking (CCRN), which is also the focus of this book.

1.2 Spectrum Sensing

Traditional, spectrum sensing is an essential component of cognitive radio to exploit the spectrum bands opportunistically. The objective of spectrum sensing is to check the channel availability in order not to adversely affect the performance of PUs [20, 21]. Since spectrum holes can be in specific time, or a frequency band, or at a spatial location, spectrum sensing can be performed in the time, frequency, and space domains. Although the main job of spectrum sensing is to obtain channel availability information, it can also be used to determine the types of signals occupying the spectrum, which may include modulation, carrier frequency, waveform, bandwidth, etc.

SUs perform spectrum sensing before commencing transmission to avoid interference to PUs. Particularly, the SU scans a certain spectrum range and detects the spectrum hole, and then accesses the channel for its transmission. During the transmission, if the SU detects the presence of PUs, it must refrain from utilizing that band and searches for a new band. In the literature, popular detection techniques include energy detection, cyclostationary detection, and matched filtering [21–24].

1.2.1 Energy Detection

Energy detection is based on the fact that the energy of the signal is usually larger than that of noise. To determine the existence of PUs, the energy detector compares its output (e.g, the average or the total energy of the observed samples) with a predefined threshold, which is derived based on the statistic of noise. If the output is above the threshold, then the energy detector makes the decision that the PU is present; otherwise, it makes the decision that the PU is absent. Energy detection is the most common type of spectrum sensing technology because of the following reasons: (1) it is simple to implement; (2) it does not require any a priori information regarding the PUs' signal; (3) the detection time is relatively short [25–28]. To evaluate the performance of the detector, the probability of detection and the probability of false alarm are used, which are defined by the probabilities that the output of the detector is above the threshold given that the PU is actually present, and the output of the detector is above the threshold when there is no PU present, respectively. As a good detector, it should have a high probability of detection and a low probability of false alarm.

1.2.2 Cyclostationary Feature Detection

Typically, there are certain inherent features associated with the signal transmitted by PUs, which can be exploited to detect the presence of PUs. Considering that for most communication systems the signals are cyclostationary due to the periodicity in the signals or the statistics, while the noise is usually assumed as a wide-sense stationary process without correlation; the cyclostationary features can be leveraged to distinguish the PUs' signal and noise [29,30]. Through the cyclostationary feature detection, features of PUs' signal can be extracted to determine the existence of PUs. Compared with energy detection, cyclostationary feature detection can provide better performance for the scenario of low SNR, with the price of high complexity. Moreover, a priori knowledge regarding the characteristics of PUs' signal is needed.

1.2.3 Matched Filter Detection

In most wireless systems, pilot bits are periodically transmitted for channel estimation, synchronization, and so on. The pilot bits are public information, which can be used to detect the presence of PUs. When the knowledge about the transmitted signal is available in the first place, matched filter detection will be the optimal detection approach, because it can correlate the received signal with the known primary signal for the detection. The match filter has the advantage of short detection time and it works well in the low SNR regime. But it requires perfect knowledge of the characteristics of PUs' signals, e.g., modulation type, bandwidth, center frequency, etc. Any imperfect information about the PU's signal will lead to severe degradation in the detection performance.

1.2.4 Limitations

Spectrum sensing is critical for the operation of CR. However, the performance of sensing is limited by several factors, including multipath fading, shadowing, primary receiver uncertainty problem [31]. When the SU is experiencing multipath fading or shadowing, the reception of PU's signal will be significantly degraded, which adversely affects the detection accuracy. In addition, for SUs which are out of the transmission range of the primary transmitter, they cannot detect the PU's transmission. Therefore, when those SUs start to transmit, harmful interference will be created at the primary receiver, if the primary receiver is unfortunately located

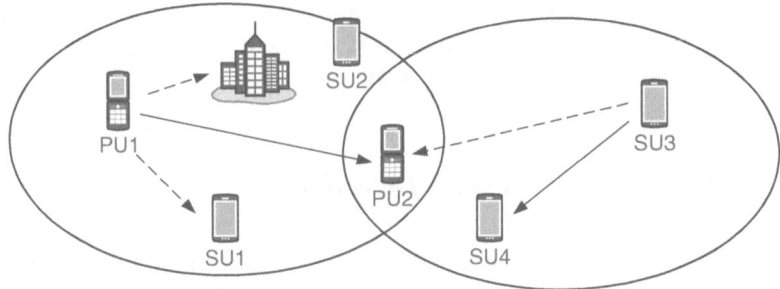

Fig. 1.5 Limitations of spectrum sensing

within the transmission range of the SUs, which gives rise to primary receiver uncertainty problem. As illustrated in Fig. 1.5, when PU_1 is transmitting data to PU_2, SU_1 can receive signal of PU_1 and know the presence of PUs. However, SU_2 cannot detect PU_1 because the building blocks the signal from PU_1. For SU_3, since it is outside of the transmission range of PU_1, it cannot detect the PU's transmission and therefore it starts its own transmission, which will cause interference to the primary receiver, i.e., PU_2. Moreover, spectrum sensing consumes energy to detect the spectrum holes and has to be continuously carried out during the transmission to detect PUs's activities.

1.3   Cooperation in CRNs

To overcome the limitations of spectrum sensing, cooperation has been introduced in CRNs, which has two forms: cooperative spectrum sensing (CSS) [32–34] and cooperative cognitive radio networking (CCRN) [35–37]. For the former, the cooperation is carried out among SUs, where multiple SUs cooperate with each other to enhance the detection performance. For the latter, the cooperation is carried out between SUs and PUs, where SUs can gain transmission opportunities through cooperation with PUs. As a promising paradigm, CCRN can relieve SUs from the burden of spectrum sensing. Moreover, it can solve the issue of dynamics of SUs' transmission caused by the PUs' activities. Therefore, we mainly focus on CCRN in this book.

1.3.1   Cooperative Sensing

Cooperative spectrum sensing that relies on spatial diversity and multiuser diversity can improve the detection performance in terms of increasing the detection probability and reducing the false-alarm probability [38, 39], as shown in Fig. 1.6. Instead of using individual decision, multiple SUs share the sensing results to make

Fig. 1.6 Cooperative sensing

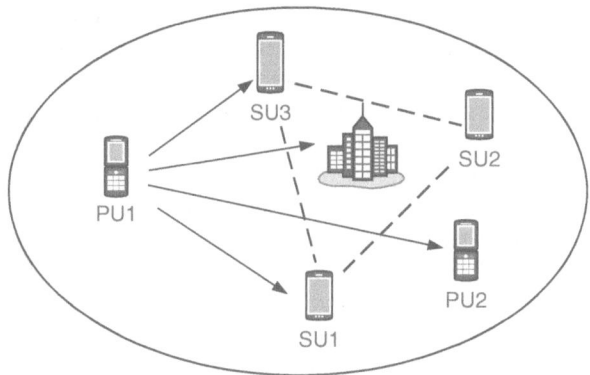

a combined decision through cooperation. Particularly, each SU performs local sensing and reports the detection results to a fusion center to make a final decision in a centralized fashion, or exchange the local detection results among themselves in a distributed fashion. Through cooperation, SUs share their sensing results and make a combined cooperative decision derived from the spatially collected observations, which can overcome the deficiency of individual observations at each SU. It has been shown that cooperative spectrum sensing can effectively combat multipath fading and shadowing, mitigate the receiver uncertainty problem, and hence significantly improve the detection performance [22, 32, 33, 40].

Typically, for the centralized cooperative spectrum sensing, it is carried out following a three-step process. First, individual SUs perform local sensing separately. Then, all the cooperating SUs forward the sensing results to the fusion center, which might be the base station, a common receiver and so on. Last, the fusion center combines all the received sensing results and makes a final decision on whether the PU is present or absent on the observed band. For the distributed cooperative spectrum sensing, where there is no fusion center, SUs exchange the detection results among themselves and then converge to a final decision after several iterations.

Many works on cooperative spectrum sensing have been reported in the literature [32, 33, 41–43]. The authors in [32] propose a cooperative spectrum sensing scheme to improve the spectrum sensing in the presence of shadowing or fading effects. In [42], the authors propose a relay-based cooperation mechanism, which is a two-user cooperative spectrum sensing scheme. This cooperation scheme shows that the detection time can be reduced. The authors in [41] propose a selective-relay based cooperative sensing scheme with no dedicated reporting channel. In [43], they also study the sensing and transmission trade-off and show that the performance in terms of the spectrum hole utilization can be significantly improved using cooperative relaying. An optimal sensing scheme for the multiuser cooperation is proposed in [33].

If all the participating SUs behave well, the detection performance can be improved through cooperative sensing. However, in an untrusted, even hostile environment, malicious users might launch different attacks to disturb the detection, hence jeopardize the operations in CRNs. For instance, malicious users might transmit signals with characteristics similar to those of PUs or just send jamming signals to the target channel to interfere with the sensing process and significantly reduce the throughput of legitimate SUs. The former is usually referred to as primary user emulation (PUE) attack, while the latter is called jamming attack. Moreover, the malicious user might send false sensing report to the fusion center, so as to mislead the spectrum sensing results and severely degrade the performance of cooperative sensing [40], which is called false sensing report attack. In [44], the authors show that the PUE attack can significantly reduce the available resources to the SUs and propose a transmitter verification scheme to combat the PUE attack. To deal with the false sensing report attack, the authors in [45] propose a robust distributed spectrum sensing to achieve a relatively accurate sensing decision in spite of that attack. While the authors in [46] propose an outlier detection to pre-filter the extreme values from sensing data. To mitigate the control channel jamming attack, the authors in [47] propose a randomized distributed scheme based on frequency hopping, which allows SUs to establish a new control channel.

1.3.2 Cooperative Networking in CRNs

1.3.2.1 Cooperative Communication

Cooperative communications have been extensively studied in the literature. The basic idea behind cooperative communication is as follows: when the source transmits message to the destination, the nodes in between can also receive it due to the broadcast nature of the wireless media. Those nodes can process the received signal and retransmit to the destination. Therefore, the destination can make use of the multiple copies of the message to create spatial diversity to improve the reception performance. It is recognized that cooperative communications can improve the transmission rate, save energy, enhance the reliability and so on. In the following, we only briefly introduce two cooperative communication approaches: Amplify-and-Forward (AF) and Decode-and-Forward (DF).

For AF relaying mode, the source first transmits the signal to the destination, which is also overheard by the relay. When the relay receives the signal from the source, it just scales the received signal by a factor, and then forwards the amplified version to the destination. After receiving those copies, the destination combines them using maximum ratio combining (MRC) to achieve the optimal reception. The overall signal-to-noise ratio (SNR) at the destination is equal to the sum of the received SNRs from both links. The advantage of AF mode is that it is simple to implement. However, the disadvantage is that the noise at the relay is also amplified and forwarded to the destination.

For DF relaying mode, different from the AF mode, the relay first decodes the received signal, re-encodes it, and then forwards to the destination. If the signal is decoded correctly, the noise component from the source node can be removed

Fig. 1.7 Cooperative
cognitive radio networking

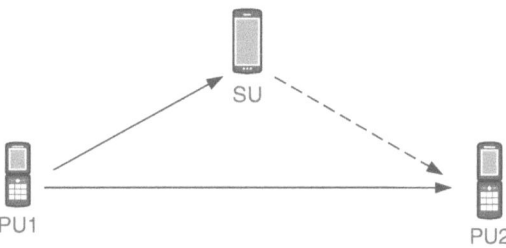

at the relay node before forwarding to the destination. If the signal is decoded incorrectly, then the relayed signal is meaningless to the destination. Therefore, the overall performance of DF mode is determined by the worst link between the link from the source to the relay and the one from the source to the destination. In addition, to further improve the performance, adaptive DF mode can be employed, which allows the relay to forward the information only when the relay decodes successfully. Compared with AF mode, DF mode has the advantage of reducing the adverse effects of noise at the relay. However, it is more complex.

1.3.2.2 CCRN

Because of the benefits of cooperative networking, there is a strong interest to introduce cooperative networking to CRNs to deal with challenges of spectrum sensing and better explore transmission opportunities. As shown in Fig. 1.7, in cooperative cognitive radio networking (CCRN), SUs cooperate with PUs to improve the latter's performance in terms of transmission rate, reliability, energy efficiency and so on, and in return gain transmission opportunities [35, 36, 48–54]. Specifically, an SU acts as a relay to improve a PU's transmission performance. Then, the PU grants a period of time to the SU as a reward, in which the SU can access the spectrum bands for transmissions. By leveraging cooperation between PUs and SUs, a "win-win" situation is created, where the PU's performance is enhanced and SUs can access the channel in the rewarding time. By this emerging cooperative networking, SUs can be relieved from the burden of spectrum sensing. The details regarding CCRN are presented in Chap. 2.

1.4 Summary

In this chapter, cognitive radio and cognitive radio networks are introduced, which are envisioned to increase spectrum utilization and solve the problem of spectrum scarcity. Spectrum sensing, as a key component of cognitive radio, is discussed, and the limitations are highlighted. Two forms of cooperation, cooperative sensing and cooperative cognitive radio networking, are introduced to solve the issues of spectrum sensing. In this book, we mainly focus on the latter form of cooperation.

Chapter 2
Cooperative Cognitive Radio Networking

Abstract As a promising paradigm, cooperative cognitive radio networking has gained prominent attentions in the literature, whereby cooperative networking is leveraged to create a win-win situation for both PUs and SUs. This chapter will first provide a comprehensive survey of existing literature in CCRN to better understand the issues related. Then, the security aspects are discussed, which are of great importance and need to be considered before the widespread deployment of cooperation in CRNs.

2.1 Literature Survey

This section will survey the state-of-the-art research in CCRN in the following categories: (1) network architecture; (2) cooperation setting; (3) cooperation on different layers; (4) communication scenario; (5) cooperation phase and (6) cooperation objective. In each category, the issues and challenges are provided, followed by a discussion on future work.

2.1.1 Network Architecture

As mentioned before, CRNs are comprised of two components: a primary network and a secondary network, either of which can be in an ad hoc mode or an infrastructured mode. Cooperation for different network architectures poses different challenges and have different features. Cooperation between two infrastructured networks is studied in [55, 56]. In [55], cooperation is studied under the scenario consisting of an infrastructured primary network and an infrastructured secondary network, where the PUs communicate with the primary base station and SUs communicate with the secondary base station. The objective of cooperation is formulated as a weighted sum throughput maximization problem, and

N. Zhang and J.W. Mark, *Security-aware Cooperation in Cognitive Radio Networks*,
SpringerBriefs in Computer Science, DOI 10.1007/978-1-4939-0413-6_2,
© The Author(s) 2014

closed-form solutions of the optimal power setting/allocation are obtained for the amplify-and-forward and decode-and-forward relaying modes, respectively. In [56], the authors take into account that both active and inactive PUs coexist in the primary network, and propose two simple cooperation frameworks that are capable of stimulating both active and inactive PUs, and SUs to participate in cooperation to achieve mutual benefits. Specifically, for active PUs, SUs can relay PUs' packets and obtains transmission opportunities as a reward. For inactive PUs, neighboring SUs can lease spectral bands together from inactive PUs, and perform cooperative communication with each other. In [57], the cooperation is considered for a scenario, where the primary network is an infrastructured network and the secondary network is an ad hoc network. Multiple SUs compete with each other to maximize their own utilities via cooperation with the PU and the PU selects the best SU with which it can achieve the maximum utility.

2.1.2   Cooperation Setting

The cooperation settings under consideration includes: (1) single PU and single SU; (2) single PU and multiple SUs; and (3) multiple PUs and multiple SUs. For single-user cooperation, the main issue is resource allocation, such as power and time allocation. For multi-user cooperation, the issues of relay selection and coordination need to be considered.

Cooperation schemes between one single PU and one single SU based on cooperative amplify-and-forward (AF) and decode-and-forward (DF) relaying are proposed respectively in [58] and [51]. In [58], AF cooperative communication is employed when the SU acts as a helper to relay the PU's signal, while in [51], the SU first decodes PU's signal and then do the forwarding to improve the transmission performance of the PU. In [59], a more general multi-user scenario is considered, where there exists one primary pair (e.g., a primary transmitter and receiver) and multiple SUs seeking transmission opportunities. A distributed secondary user selection scheme is proposed, which optimizes the performance of the secondary system without degrading the performance of the primary system. Specifically, the SU, which can help the PU to achieve the targeted transmission rate, is selected to relay the PU's message. In the meanwhile, the SU which can minimize the outage probability of the secondary network is selected to transmit simultaneously, under the constraint that its interference should not be above the interference threshold of the PU.

In [35, 36], the cooperation between a single PU and multiple SUs is studied by using Stackelberg game, whereby the PU cooperates with a set of SUs to increase PU's transmission rate [36], or to improve the PU's utility in terms of the transmission rate and the revenue obtained from SUs [35]. In the rewarding time, multiple SUs share the spectrum granted by the PU, either using a power allocation game [36] or a payment mechanism [35]. A similar scenario can be found in [54], where multiple SUs compete to cooperate with the PU by acting as a relaying node to gain the transmission opportunities. The relaying SU is selected among multiple candidates using an auction mechanism, where the PU, the competing SUs and the access time slot are modeled as the auctioneer, the bidders and the bidding article, respectively. In [49], the authors extend the framework of CCRN by considering the presence of multiple PUs. Multiple PUs compete with each other for cooperation with SUs which have QoS requirements on spectral resources received for their own transmission. The scenario is modeled as a generalized Nash equilibrium (GNE) problem, and the GNE and variational inequality solutions are discussed.

In addition, the authors in [37, 52] consider the existence of multiple PUs performing cooperation with multiple SUs in the network. Specifically, in [52], the transmission of PUs are divided into different frames and different pairs of PU and SU perform cooperation over different frames to maximize the network utility. In [37], cooperation for multi-channel CRNs is investigated, where multiple PUs operating over different channels cooperate with different SUs simultaneously to maximize the network utility. To this end, the maximum weight matching is utilized to coordinate the cooperation between multiple PUs and multiple SUs.

2.1.3 Cooperation on Different Layers

The cooperation between PUs and SUs can be performed on different layers, such as physical layer, link layer and network layer. For physical layer, the cooperation objective involves the outage probability, transmission rate, and so on. Moreover, diverse physical layer techniques can be leveraged to facilitate cooperation, such as superposition coding, network coding, quadrature signalling. For the link layer, the cooperation takes advantages of the retransmission scheme (ARQ) to exploit cooperation benefits. For the network layer, cooperation is carried out to find the optimal routing and so on.

For cooperation on physical layer, superposition coding and network coding are considered in [60], where two PUs communicate with each other with the help of an SU. The SU first decodes the signals from the two PUs, then employ network coding to encode them. After that, the SU superposes its own signal with the network-coded primary signals and broadcast using different power levels. The outage probabilities for both the primary system and the secondary system under the proposed cooperation scheme are derived. Moreover, the cooperation range is obtained, and once the SU is located in that range, the cooperation is beneficial to both the primary system and secondary system.

In [61, 62], quadrature signaling is utilized during cooperation between PUs and SUs. In [61], a cooperation scheme based on quadrature signaling is proposed, whereby the PU and the cooperating SU use quadrature amplitude modulation (QAM) to attain orthogonal signaling to cooperate efficiently. The SU selects the optimal power allocation coefficient to maximize the performance of the PU when its own transmission requirement is satisfied. In [62], quadrature signaling is utilized for SUs to share the spectrum received from the PU. In [50], the capability of multiple antennas is exploited to facilitate cooperation, where SUs are equipped with multiple antennas and PUs are equipped with single antenna. When PUs are transmitting, SUs can receive signals from another SU at the same time if the total number of on-going traffic stream is no more than the Degree-of-Freedom (DoF) provided by SUs' antennas. When SUs are forwarding PUs' signal, SUs can simultaneously transmit their own signals using beamforming to mitigate interference to the primary receivers.

Cooperation on link layer is investigated in [54,63–65], which take advantage of opportunities that arise during Automatic Repeat reQuest (ARQ) retransmission. In [54], cooperative ARQ is integrated into the cooperation scheme, whereby a relaying SU is employed if needed and it is also rewarded to use a part of the time slot for its own transmission. The relaying SU is selected from multiple candidates using an auction game. A cooperation scheme is proposed in [63] for the SU to exploit transmission opportunities in the ARQ based primary system without degrading the PUs' performance. Specifically, there are two modes for the SU: cooperation mode and access mode. For cooperation mode, the SU acts as a relay to help the PUs' transmission and accumulate credits for use in the future. For access mode, the SU transmits simultaneously when the PU is retransmitting its message. To make sure that the primary system achieves higher throughput on average, the credits obtained from the cooperation mode should be sufficient to compensate the performance degradation of the primary system in access mode. In [65], the SU overhears the ACK/NACK feedback sent from the primary receiver, and then decides to access the spectrum or not. An opportunistic sharing scheme is proposed to exploit four kinds of spectrum opportunities based on the ACK/NACK of PUs.

Cooperation on network layer is studied in [66], where there exist a primary multi-hop network and a set of SUs. Opportunistic routing is employed to improve the throughput of the primary network over fading channels, which selects the next hop in an opportunistic way according to the decoding outcomes of the previous transmission. To exchange for spectrum access opportunities, the SUs can serve as potential next hops that route packets based on opportunistic routing.

2.1.4 Cooperation Phases

In terms of cooperation phases, all existing work can be divided into two groups: two-phase cooperation schemes and three-phase cooperation schemes. For the three-phase cooperation, the PU transmits in the first phase, then the SUs relay the PU's signal in the second phase, and in the last phase the SUs access the spectrum as a reward. In the literature, there are ample works that make this type of scheme more efficient and realistic. On the other hand, different physical layer techniques, e.g., quadrature signalling, beamforming, superposition coding, network coding, are leveraged to merge the last two phases to facilitate two-phase cooperation schemes.

In [36], a three-phase cooperation scheme is proposed, whereby the PU transmits message to a set of SUs in the first phase, the SUs which decode the message successfully relay the PU's message via distributed space-time coding (DSTC). In the last phase, the cooperating SUs start their transmissions by selecting suitable power levels. In [35], the cooperation between PUs and SUs is also performed in a three-phase fashion, whereby SUs cooperate with the PU to improve the PU's utility and then share the rewarding resource via a payment mechanism.

A two-phase cooperation scheme is proposed in [51], whereby the PU transmits its signal to the SU in the first phase, and then the SU decodes the received signal and superimposes it with its own signal to broadcast in the second phase, using different power levels. In [55], the authors propose a two-phase cooperation scheme, whereby the PU selects an SU for cooperation to maximize the throughput, and the selected SU relays the PU's signal and transmits its own signal simultaneously by leveraging the degrees of freedom provided by orthogonal modulation. In [67], a novel polarization enabled two-phase cooperation framework for cognitive radio networking is proposed. By leveraging the degrees of freedom provided by orthogonally dual-polarized antennas, secondary users can relay the traffic of PUs and transmit their own traffic in the same time slot without interference.

In [58], a two-phase scheme based on cooperative AF relaying protocol is proposed. The key feature of the proposed approach is that the SU linearly combines the primary signal with the secondary signal and allocates fractions λ and $1 - \lambda$ of the transmission power to the primary and secondary signals separately. It is shown that for a fixed λ, a critical region can be found where the outage probability of the primary system is kept lower than the case without spectrum sharing.

In [68], network and superposition coding are utilized to facilitate two-phase cooperation for the case where multiple primary channels exist. Two PUs transmit data to a common destination with the assistance of an SU that acts as a relay and the SU attains the opportunity of transmission in return. Network coding and superposition coding are utilized at the relay SU to superimpose its own message on network-coded primary signals. The probabilities of outage regions are obtained for multiple access channels.

2.1.5 Cooperation Objective

The cooperation objective for PUs and SUs could be the same or different, which has different approaches to model the interactions between PUs and SUs. In the literature, the objectives of PUs include transmission rate, reliability, energy saving, security, connectivity, and so on.

In [35, 36, 49, 53], the objective of cooperation is to maximize the PU's transmission rate and provide transmission opportunities to SUs. In [36], the cooperating SUs relay the PU's codeword via DSTC and share the rewarding resource using a power allocation game. In [35], SUs cooperate with the PU to improve the PU's utility, which is a combination of transmission rate and the revenue obtained from SUs. Then, multiple cooperating SUs share the rewarding resource via a payment mechanism.

In [54, 66], cooperation is performed by considering the reliability of the PU. In [54], a novel distributed scheme that integrates cooperative ARQ in the cooperation scheme is proposed. The SU which successfully decodes the PU's message can act as a relay to increase the reliability of primary transmission, and in return gain a certain period of time for its own transmission.

In [57, 69], the energy efficiency of the primary system is considered. In [69], the authors propose a frequency-division multiple access (FDMA) based two-phase cooperation in which a PU divides its spectrum into two orthogonal subbands and broadcasts on the first subband in the first phase. SUs relay the PU's signal on the same subband in the second phase, and continuously transmit in both phases on the second subband. In [57], a cooperation scheme based on time-division multiple access (TDMA) is proposed, where an SU cooperates with a PU to improve the energy saving of the primary transmission and gains transmission opportunities. As the license holder, the PU can decide when to cooperate, with whom to cooperate, and how to cooperate.

In [70–72], cooperation schemes are studied, which improve the physical layer security of the primary link and provide transmission opportunities to SUs. The security level is measured by secrecy rate, which is defined by the difference between the transmission rate at the primary receiver and that at the eavesdropper. The objective of cooperation is to maximize the secrecy rate of the primary transmission.

2.2 Security Aspects

Cooperation can bring many benefits if all nodes are well-behaved. However, in reality, this assumption may not hold. When there exist some dishonest or malicious SUs, cooperation can incur security issues, which will compromise the cooperation benefits and disturb the normal operation of CCRN. On the other hand, since cooperation can be leveraged to increase the secrecy of transmission, security also brings opportunities for cooperation.

2.2.1 Trust-Aware Cooperation

In CCRN, the PU's packets can be eavesdropped by the neighboring SUs such that the confidentiality can not be guaranteed. Moreover, the malicious relays can alter the PU's packets or fabricate its packets and then forward them to the destination. To secure cooperation between PUs and SUs, the following basic security requirements need to be met: confidentiality, integrity, and authentication, which can be provided by suitable cryptographic approaches (e.g., encryption and decryption, authentication, message authentication code, digital signature, etc.). However, a legitimate SU may be compromised and misbehaves during cooperation when it is selected to cooperate with the PU, e.g., it may launch black or grey hole attack [73]. A dishonest SU may not obey the cooperation rule during the cooperative transmission to pursue more self-benefits, e.g., it may transmit its own packets instead of relaying the PU's packets. Furthermore, considering the mobility of SUs, the malicious or dishonest SUs may misbehave at one place then move to other places. Since there is no record of the past behaviors, these users can have the same opportunity to participate in cooperation with the PU, and then continue to harm the system. In a nutshell, without considering these security threats, the PU may choose an untrustworthy SU for cooperation, which will cause the failure of the cooperation and degrade the PU's quality of service (QoS).

As a summary, the potential misbehaviors in CCRN can be listed as follows.

1. Selfishness: the cooperating SU may choose a lower transmission power than the expected one during cooperation or it just chooses not to forward the PU's message to save energy.
2. Maliciousness: the malicious SU may delete, modify or replace the bits in the DF mode. In AF mode, it may intentionally add some jamming signals to corrupt the PU's signal.
3. Dishonesty: the dishonest SU may present the fake CSI to gain transmission opportunities.

Therefore, the security issues need to be considered, when PUs choose to cooperate with SUs.

2.2.2 Cooperation for Secrecy

As mentioned above, cooperation can incur diverse security issues. In the meanwhile, security also brings opportunities for cooperation. Specifically, in a hostile environment, there exist some unfriendly users, e.g., eavesdroppers. Due to the broadcast nature of wireless communication, these unfriendly users can easily overhear the ongoing transmission. This consequence not only hurts the confidentiality of communications, but also exposes the vulnerabilities that a malicious user can exploit to launch attacks to the network. To protect the secrecy of transmissions, the traditional solution is to use encryption at upper layers of the communication protocol. However, the security scheme at upper layers is prone to potential attacks. Moreover, it becomes very challenging for a network without infrastructure [74]. In addition, in a network where the nodes have relatively low power, e.g., sensors, it might not be practical to implement cryptographic algorithms [74]. To secure the communication effectively, there is a novel approach at the physical (PHY) layer [74–76], which exploits the characteristics of the wireless channel for secure transmission. In [75], it is shown that perfectly secure information can be exchanged at a nonzero rate between the source and destination. However, it becomes infeasible when the source-destination channel is worse than the source-eavesdropper channel. In [77], the authors studied secure communications in the low SNR regime, which corresponds to the cases of long distance transmissions or energy-limited scenarios. The problem is that all the nodes are assumed to be equipped with multiple antennas, which may be infeasible in reality. To address the above issues, user cooperation has been introduced to enhance the secrecy of communications [78]. In the context of CCRN, a cooperation based spectrum access has been proposed in [70], which improves the security of the primary link and provide transmission opportunities to SUs. Nevertheless, it is achieved at the expense of employing multiple antennas as well. Observing the above, we are inspired to design cooperation schemes for CRNs, where all users are only equipped with single antenna to enhance the security of the primary link and provide transmission opportunities to SUs.

2.3 Summary

In this chapter, we focus on cooperative cognitive radio networking and a literature review is provided, where cooperation schemes are classified into different categories. The security aspects of this cooperative paradigm and the limitations of the existing works are discussed. In the subsequent chapters, security-aware cooperation schemes will be studied.

Chapter 3
Trust-Aware Cooperative Networking

Abstract In this chapter, we introduce a trust-aware cooperation scheme to facilitate the cooperation and SU selection in an unfriendly environment, whereby the primary users (PUs) choose trustworthy partners as relays to improve throughput or energy efficiency. Specifically, the cooperation involves a PU selecting the most suitable secondary user (SU) to relay its message and giving the selected SU spectrum access right as a reward, taking the trustworthiness of SUs into consideration. While the SU, being starved for transmission opportunities, chooses a suitable power level for cooperation and its own transmission. The proposed cooperative strategy, including partner selection, time slot allocation and power allocation in an unfriendly environment, are analyzed. Numerical results demonstrate that, with the proposed scheme, the PU can achieve higher throughput or energy saving through cooperation with the trustworthy SU.

3.1 Introduction

Cooperative networking can be leveraged to deal with the limitations of spectrum sensing [31], whereby SUs cooperate with PUs to improve the PUs' transmission performance, and in return gain transmission opportunities. Therefore, both PUs and SUs can benefit from cooperation, which creates a win-win situation. In the literature, this paradigm is referred to as cooperative cognitive radio networking (CCRN) [35–37, 48–51, 54]. The common assumption for the existing works is that SUs are trustworthy and well-behaved during cooperation, which may not be always true in reality. In an unfriendly environment, where there exist selfish, even malicious SUs, security issues mentioned in Chap. 2 can arise, which will compromise the normal operation of CCRN. Thus, security needs to be considered for this emerging cooperative networking, which has not been given due consideration in the literature.

In this chapter, we make an effort to guarantee and improve the performance of cooperation and SU selection in an unfriendly environment, where SUs may

N. Zhang and J.W. Mark, *Security-aware Cooperation in Cognitive Radio Networks*,
SpringerBriefs in Computer Science, DOI 10.1007/978-1-4939-0413-6_3,
© The Author(s) 2014

misbehave during cooperation. Moreover, cooperation under two scenarios with different primary link qualities are considered, where the PU may have different concerns. In particular, if the primary link is poor so that the transmission quality drops dramatically, the PU has the incentive to increase the throughput via cooperation with SUs so that the QoS requirement can be satisfied; if the channel condition is good enough, since the PU can meet the traffic demand on its own, it may have a concern with energy efficiency due to power limitation. By the same token, the SU may value the throughput more since it typically does not have much transmission opportunities. Based on the fact that the PU and the SU have different incentives for cooperation, we propose a cooperative spectrum access scheme which also incorporates trust value into the system to mitigate the security issues. Considering that PUs and SUs aim at maximizing their own utilities, we leverage game theory to analyze the cooperation scheme. Furthermore, as the license holders, the PUs have higher priority on spectrum usage and are supposed to lead the cooperation. Thus, we model the interaction between the PU and the SU as a Stackelberg game [79], which provides the PU with the best strategy in the partner selection process, spectrum access time allocation and transmission power determination. Based on the analysis of the game, the PU selects the best SU and optimal cooperation parameters to maximize the utility if cooperation is agreed upon. That is, in accordance with the outcomes of the game, the PU can decide when to cooperate, with whom to cooperate and how to cooperate.

3.2   System Model

This section presents the details of system model and the main system parameters.

3.2.1   MAC Layer

As shown in Fig. 3.1, there exist two types of networks, the infrastructure-based primary network and ad hoc secondary network, collocated in the same area. In the primary network, the PUs communicate with the base station (BS) in time-division multiple access (TDMA) mode, while the SU transmits data to its corresponding receiver in the secondary network. The time slot duration for each PU's transmission is denoted by T. For cooperation, the PU selects one SU as a cooperating relay and Amplify-and-Forward (AF) protocol is employed. The PU grants the use of the bandwidth to the cooperating SU so as to improve the energy efficiency of the communication to the BS. Specifically, a fraction α of T $(0 < \alpha \leq 1)$ is used for the primary and cooperative transmission. The PU transmits data to the SU in the first $\frac{\alpha T}{2}$, which can be also overheard by the BS (Fig. 3.1a). In the subsequent duration of $\frac{\alpha T}{2}$, the SU relays the received data to the BS (Fig. 3.1b). In the remaining

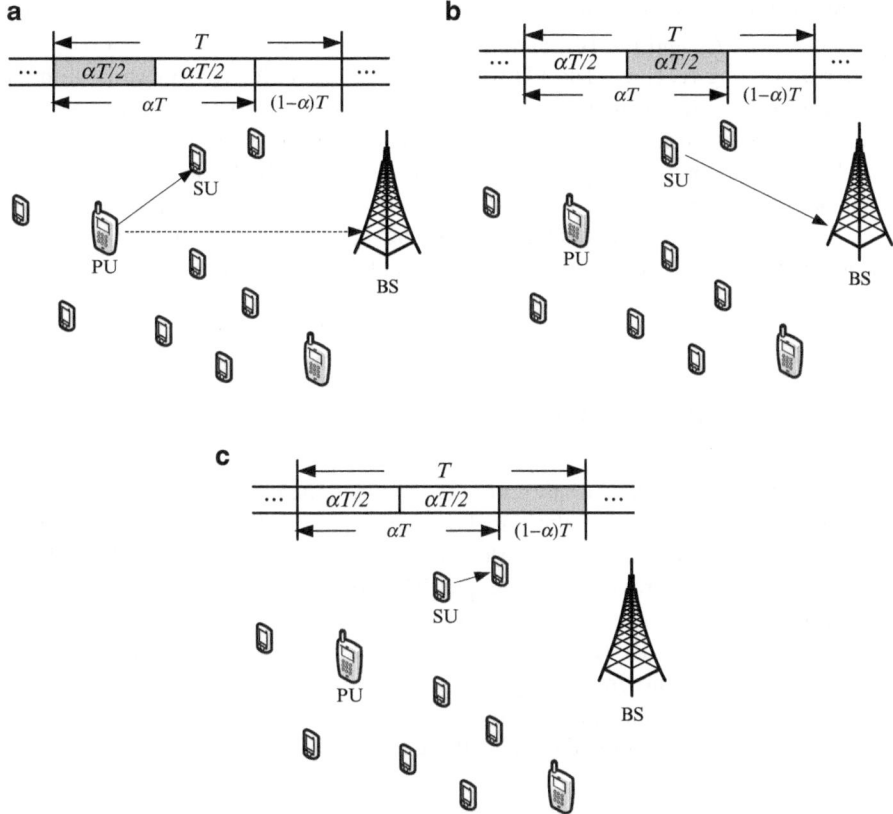

Fig. 3.1 The proposed framework of cooperation in CRNs. (**a**) Primary transmission, (**b**) cooperative transmission, (**c**) secondary transmission

duration of $(1 - \alpha) \cdot T$, the cooperating SU is allowed to transmit its own data to the corresponding secondary receiver (Fig. 3.1c). A common control channel is considered to be available for exchanging information among PU, SU and BS (e.g., channel state information (CSI), trust values, etc.) and for delivering the decision of the PU (e.g., slot allocation, the selected SU, etc.) to the secondary network.

3.2.2   Physical Layer

The channels between nodes can be modeled as independent proper complex Gaussian random variables, constant within each slot, but generally varying over the slots. We use the following notations to denote the instantaneous channels in each slot: h_{pb} denotes the complex channel gain between the PU and the BS;

h^i_{ps} denotes the channel gain between the PU and SU_i; h^i_{sb} denotes the channel gain between SU_i and the BS; and h^i_s denotes the channel gain between SU_i and its corresponding receiver. The PU uses power P_d watts for transmission without cooperation. As shown in Fig. 3.1, with cooperation, the PU chooses power P_c watts for primary transmission. The bandwidth owned by the primary user is W Hz. SU_i is constrained to spend the same power P^i_s for both the cooperative and secondary transmission so as to ensure SU_i spend at least the same power which it is willing to spend for its own transmission. The one-sided power spectral density of the independent additive white Gaussian noise at the both base station and secondary receivers is N_0.

3.3   Cooperation for Throughput Improvement

In this section, we will discuss the cooperation when the primary link is weak. Due to the poor channel condition, the signal reception quality at the destination receiver will be degraded dramatically. In order to meet the traffic demand, the PU selects a suitable relay to improve the throughput. The objective of the PU is to maximize the throughput through cooperation in an unfriendly environment. To evaluate the risks of cooperation, trust value is applied and the cooperation procedure is modeled using Stackelberg game. In such a game, the utilities of both the PU and the SU are presented and analyzed; and the close-form solutions for the players' best strategies are derived, which constitute the Stackelberg equilibrium.

3.3.1   Trust Computational Model

In an unfriendly environment, the aforementioned security issues can rise, which cannot be well mitigated by means of cryptographic methodologies [80]. Thus, trust and reputation system is applied to address these issues [81]. Specifically, trust values are assigned to SUs and utilized to evaluate the behaviors of SUs. The PU maintains a table for recording identities and the corresponding trust values of its one-hop neighboring SUs. In addition, BS keeps the trust values of all SUs in its domain. Each time after cooperation, the behavior of the selected SU will be evaluated and the trust value will be updated accordingly. Then, the trust value will be exchanged periodically between the PUs and the BS.

We use a Bayesian framework [82, 83] to evaluate the trust values: each entity is assumed to behave well with probability p, and misbehave with probability $(1 - p)$, i.e., the behavior of the entity follows a Bernoulli distribution. Through a series of observations, a posteriori probability can be derived to estimate the future behaviors of the entity. Posteriori probabilities of binary events can be represented as the beta distribution. An expression of the probability density function (PDF) $f(\hat{p}|\kappa,\iota)$ in terms of the gamma function Γ is given by:

$$f(\hat{p}|\kappa,\iota) = \frac{\Gamma(\kappa+\iota)}{\Gamma(\kappa)\cdot\Gamma(\iota)} \cdot \hat{p}^{(\kappa-1)} \cdot (1-\hat{p})^{(\iota-1)}, \tag{3.1}$$

where $\hat{p}$ is the estimate of p, and κ, ι are the two parameters. The expectation of beta distribution is given by $E(\hat{p}) = \frac{\kappa}{(\kappa+\iota)}$, which can be used to represent the trust value of the relevant entity.

In our system, a malicious or dishonest SU_i behaves well with probability p_i and misbehaves with probability $1 - p_i$. In order to estimate the trustworthiness of SUs, BS needs to observe the ongoing transmission and evaluate the activities of SUs according to the received signals. To determine whether the relaying SU misbehaves or not, one approach is to utilize tracing symbols, which are known at both the source and the destination [84,85]. Another way is based on the correlation between signals received from the source and the relay [86]. In addition, the misbehavior can also be detected based on the success or failure of transmitted frames via acknowledgment (ACK/NACK) [87]. Based on the related work in the literature, the misbehavior of relaying nodes can be detected, which is beyond the scope of this work. Consider a process with two possible outcomes (misbehavior or well-behavior) and let μ be the observed number of good behaviors and ν be the observed number of misbehaviors. Then, the PDF of observing outcomes in the future can be expressed as a function of past observations by setting: $\kappa = \mu + 1$ and $\iota = \nu + 1$. Thus, the expected value of $\hat{p}$ can be determined from observations as follows:

$$E(\hat{p}) = \frac{\mu+1}{(\mu+\nu+2)}, \tag{3.2}$$

which is used as the trust value Tr_i of SU_i.

When new observations of a particular SU are made, e.g., δ observed misbehaviors and ξ observed good behaviors, the associated trust value can be updated using (3.2) by setting $\nu := \nu + \delta$ and $\mu := \mu + \xi$.

3.3.2 Stackelberg Game between PU and SU

Stackelberg game is applied to model the cooperation procedure, considering different priorities for spectrum usage of the primary system and the secondary system. In the Stackelberg game, the PU acts as the leader and the SU is the follower. As the leader, the PU can choose the best strategies, aware of the effect of its decision on the strategies of the follower (the SU); while the SU can just choose its own strategies given the selected parameters of the PU. The utility functions for both PU and SU are respectively defined. By analyzing the game, the best cooperating SU and the optimal cooperation parameters can be determined.

3.3.2.1   Primary User

Under the scenario of poor channel condition, the throughput of direct transmission is reduced. Thus, the PU is interested in cooperation to increase the throughput. Given a fixed time duration T, increasing the throughput is equivalent to increasing the average transmission rate. To this end, the PU selects the most suitable SU from the set $\mathbf{S}_p$ of its one-hop neighbors. Suppose SU_i is chosen for cooperation, the PU decides the time allocation parameter α_i and its transmission power P_c^i to maximize throughput, on the basis of available instantaneous CSI. The average transmission rate R_c^i through AF cooperative communication between the PU and SU_i is given as follows:

$$R_c^i = \frac{\alpha_i W}{2} \log_2 \left[1 + \frac{P_c^i \left| h_{pb} \right|^2}{N_0} + f\left(P_c^i \left| h_{ps}^i \right|^2, P_s^i \left| h_{sb}^i \right|^2 \right) \right], \qquad (3.3)$$

where

$$f\left(P_c^i \left| h_{ps}^i \right|^2, P_s^i \left| h_{sb}^i \right|^2 \right) = \frac{1}{N_0} \frac{P_c^i \left| h_{ps}^i \right|^2 P_s^i \left| h_{sb}^i \right|^2}{P_c^i \left| h_{ps}^i \right|^2 + P_s^i \left| h_{sb}^i \right|^2 + N_0}.$$

The factor $\frac{\alpha_i}{2}$ accounts for the fact that $\alpha_i T$ is used for cooperative relaying, which is further split into two phases. Considering the trust value Tr_i of each neighboring SU_i, the utility function is given by

$$U_p^i = Tr_i \cdot R_c^i, \qquad (3.4)$$

which indicates the expected throughput the PU can attain through cooperation with SU_i. The objective of the PU is to maximize its utility function and the strategy is to choose the most suitable SU from the set of its one-hop neighboring SUs and the cooperation parameters, i.e., the time allocation parameters α_i and the transmission power P_c^i for cooperation with the selected SU_i.

3.3.2.2   Secondary User

The SU can gain transmission opportunities through cooperation with the PU. In particular, the SU relays PU's data in the second phase and transmits its own data in the last phase. Assuming cooperation with the PU, the selected SU_i decides its transmission power, pertaining to the given α and P_c. The target of the SU is to maximize throughput (equivalent to the transmission rate) without expending too much energy. Following the cooperation agreement, SU_i spends the same power P_s^i for both cooperative and secondary transmissions. In particular, the transmission rate R_s^i for secondary transmission (Fig. 3.1c) between SU_i and its corresponding receiver is given by

$$R_s^i(\alpha_i) = (1 - \alpha_i)W \log_2\left(1 + \frac{P_s^i \left|h_s^i\right|^2}{N_0}\right).$$ (3.5)

With energy consumption $P_s^i(1 - \frac{\alpha_i}{2})T$, the utility function of SU_i can be represented by $R_s^i(\alpha_i)T - c \cdot P_s^i(1 - \frac{\alpha_i}{2})T$, where c ($0 < c < 1$) is the weight of energy consumption in the overall utility. With a smaller c, the SU is more concerned with throughput than energy consumption, and vice versa. Over the period of T, the utility function of SU_i is given by

$$U_s^i(\alpha_i) = W \log_2\left(1 + \frac{P_s^i \left|h_s^i\right|^2}{N_0}\right)(1 - \alpha_i) - c\left(1 - \frac{\alpha_i}{2}\right)P_s^i.$$ (3.6)

The objective of SU_i in the game is to maximize its utility by choosing the optimal transmission power P_s^i.

3.3.3 Game Analysis

As a sequential game, the Stackelberg game can be analyzed by the backward induction method. First, the optimal strategy of the SU (the follower) is analyzed, assuming the strategy of the PU (the leader) is fixed. Second, the PU decides the optimal strategy, aware of the effect of its decision on the SU. By doing so, the best response functions of both the PU and the SU are derived, such that the two players can choose to maximize the corresponding utilities. Then, the Stackelberg equilibrium of the proposed game can be achieved based on the best response functions.

3.3.3.1 Best Response Function of SU

Assuming that the PU uses α_i as a basis for cooperation, SU_i maximizes its utility by selecting the optimal transmission power. Given α_i, SU_i selects the optimal transmission power, which can be formulated as the following optimization problem:

$$\max_{P_s^i} U_s^i(\alpha_i) = (1 - \alpha_i)W \log_2\left(1 + \frac{P_s^i \left|h_s^i\right|^2}{N_0}\right) - c\left(1 - \frac{\alpha_i}{2}\right)P_s^i$$

$$\text{s.t. } 0 \le P_s^i \le P_{max},$$

where P_{max} is the power constraint for SU_i. Solving the above problem, the optimal transmission power can be determined.

Let $P_s^{*i}(\alpha_i)$ be the best response function of the secondary user if the utility of SU_i can achieve the maximum value when $P_s^{*i}(\alpha_i)$ is selected, for any given α_i, i,e., $\forall 0 < \alpha_i < 1, U_s^i(P_s^{*i}(\alpha_i), \alpha_i) \geq U_s^i(P_s^i(\alpha_i), \alpha_i)$.

Theorem 3.1. *The best response function $P_s^{*i}(\alpha_i)$ of the secondary user is given by the following equation, given that the primary user selects a certain α_i for cooperation.* $P_s^{*i}(\alpha_i) = \min\{\frac{(1-\alpha_i)W}{c(1-\frac{\alpha_i}{2})\ln 2} - \frac{N_0}{|h_s^i|^2}, P_{max}\}$

Proof. Given the time allocation coefficient α_i, the utility function of SU_i is

$$U_s^i(\alpha_i) = (1-\alpha_i)W \log_2\left(1 + \frac{P_s^i |h_s^i|^2}{N_0}\right) - c\left(1 - \frac{\alpha_i}{2}\right) P_s^i. \qquad (3.7)$$

From the above equation, it is easy to prove that the utility function first increases, and then decreases with the increase of P_s^i without considering the power constraint. Therefore, there exists an optimal power such that the U_s^i can reach the maximum value. Taking the first order partial derivative of the utility function with respect to P_s^i yields

$$\frac{\partial U_s^i}{\partial P_s^i} = \frac{(1-\alpha_i)W |h_s^i|^2}{\left(1 + \frac{P_s^i h_s^{i2}}{N_0}\right) N_0 \ln 2} - c\left(1 - \frac{\alpha_i}{2}\right). \qquad (3.8)$$

Setting $\frac{\partial(U_s^i)}{\partial(P_s^i)} = 0$ yields the optimal transmission power given by

$$\frac{(1-\alpha_i)W}{c(1-\frac{\alpha_i}{2})\ln 2} - \frac{N_0}{|h_s^i|^2}. \qquad (3.9)$$

Taking the power constraint into consideration, the best response function $P_s^{*i}(\alpha_i)$ will be

$$P_s^{*i}(\alpha_i) = \min\left\{\frac{(1-\alpha_i)W}{c(1-\frac{\alpha_i}{2})\ln 2} - \frac{N_0}{|h_s^i|^2}, P_{max}\right\}. \qquad (3.10)$$

This completes the proof.   □

The first order derivative of the best response function with respect to α_i is given by $\frac{-\alpha_i W}{(-2+a)^2 C \ln 2}$, which is negative. Therefore, the optimal transmission power of SU_i is a decreasing function with the increase of α_i. It is explained by that the SU is willing to spend more transmission power during cooperation if the PU allocates more time for the SU's transmission.

3.3.3.2 Best Response Function of PU

Aware of the best response function of the SU, the PU decides its own best strategy for utility maximization. The best response function can be derived by solving the following optimization problem:

$$\max_{\alpha_i, P_c^i, i} \frac{\alpha_i W}{2} \log_2 \left[1 + \frac{P_c^i |h_{pb}|^2}{N_0} + f\left(P_c^i \left|h_{ps}^i\right|^2, P_s^i \left|h_{sb}^i\right|^2 \right) \right]$$

$$\text{s.t. } 0 < P_c^i \leq P_{max}, 0 < \alpha_i \leq 1, SU_i \subseteq \mathbf{S}_p.$$

Let $\alpha^*, P_c^{*i^*}, i^*$ be associated with the best response function of the primary user if the utility of the PU can achieve the maximum value when this strategy is selected.

Theorem 3.2. *The best response function of the primary user is with $\alpha^*, P_c^{*i^*}, i^*$, given by $(\alpha^*, P_c^{*i^*}, i^*) = \arg\max_{\alpha_i, P_c^i, i} U_p^i$. In particular, $i^* = \arg\max U_p^i$ (P_c^{*i}, α_i^*), where*

$$P_c^{*i} = P_{max}$$

$$\alpha_i^* = \begin{cases} 1 - \sqrt{1 - \dfrac{2W\left|h_{sb}^i\right|^2}{P_{max}\left|h_{ps}^i\right|^2 c + 2W\left|h_{sb}^i\right|^2 + N_0 c}}, & \text{if } \dfrac{W}{c\ln 2} - \dfrac{N_0}{\left|h_s^i\right|^2} < P_{max} \\[2em] \max\left\{ 2 + \dfrac{2}{\frac{c\ln 2}{W}\left(P_{max} + \frac{N_0}{\left|h_s^i\right|^2}\right) - 2}, 1 - \sqrt{1 - \dfrac{2W\left|h_{sb}^i\right|^2}{P_{max}\left|h_{ps}^i\right|^2 c + 2W\left|h_{sb}^i\right|^2 + N_0 c}} \right\}, & \text{otherwise} \end{cases}$$

$$\tag{3.11}$$

P_c^{*i} *and* α_i^* *are the optimal transmission power and time allocation coefficient respectively, assuming cooperation with SU_i. The optimal $P_c^{*i^*}$ and α^* correspond to the selected i^*.*

Proof. Since the first order derivative of the utility function with respect to P_c^i is always positive, U_p is a monotonically increasing function of P_c^i. Moreover, considering the parameters P_c^i and α_i are independent, P_c^i should be selected as the maximum power so that the utility can reach the maximum value. Therefore, to solve the optimization problem, it is equivalent to optimize the utility function when $P_c^i = P_{max}$ and SU_i selects the best response $P_s^{*i}(\alpha_i)$. Since the first term in (3.10) monotonically decreases with respect to α_i, its maximum value is $\frac{W}{c\ln 2} - \frac{N_0}{\left|h_s^i\right|^2}$.

When $\frac{W}{c\ln 2} - \frac{N_0}{\left|h_s^i\right|^2} < P_{max}$, $P_s^{*i}(\alpha_i)$ takes the value of the first term in (3.10). Substituting $P_c^i = P_{max}$ and $P_s^{*i}(\alpha_i)$ into the utility function of PU, the utility can be expressed by

$$U_p = \frac{\alpha_i W}{2} \log_2 \left[1 + \frac{P_{max}\left|h_{pb}\right|^2}{N_0} + f\left(P_{max}\left|h_{ps}^i\right|^2, P_s^{*i}(\alpha_i)\left|h_{sb}^i\right|^2 \right) \right], \tag{3.12}$$

which is a function of α_i. The first order derivative of (3.12) is given by

$$\frac{\partial U_p}{\partial \alpha_i} = A \cdot \alpha_i^2 + B \cdot \alpha_i + C, \tag{3.13}$$

where

$$A = P_{max} \left| h_{ps}^i \right|^2 c + 2W \left| h_{sb}^i \right|^2 + N_0 c$$

$$B = -2P_{max} \left| h_{ps}^i \right|^2 c - 4W \left| h_{sb}^i \right|^2 - 2N_0 c = -2 \cdot A$$

$$C = 2W \left| h_{sb}^i \right|^2.$$

To find the optimal α_i^* such that U_p can be maximized, set the first order derivative of (3.12) equal to 0. It can be easily found that $C < A$; then $B^2 - 4AC > 0$. Thus, the above quadratic function has real root(s). Considering the range of α_i ($0 < \alpha_i < 1$), there exists one and only one root α_r. Thus, the optimal α_i^* is given by

$$\alpha_i^* = \alpha_r = 1 - \sqrt{1 - \frac{C}{A}}$$

$$= 1 - \sqrt{1 - \frac{2W \left| h_{sb}^i \right|^2}{P_{max} \left| h_{ps}^i \right|^2 c + 2W \left| h_{sb}^i \right|^2 + N_0 c}}. \tag{3.14}$$

When $\frac{W}{c \ln 2} - \frac{N_0}{\left| h_s^i \right|^2} \geq P_{max}$, there exists α_0 in the range from 0 to 1, such that $P_s^*(\alpha_0) = P_{max}$. Specifically, $\alpha_0 = 2 + \frac{2}{D-2}$, where $D = \frac{c \ln 2}{W}(P_{max} + \frac{N_0}{\left| h_s^i \right|^2})$. The reason is that the range of D is from 0 to 1 due to the assumption that $\frac{W}{c \ln 2} - \frac{N_0}{\left| h_s^i \right|^2} \geq P_{max}$. Therefore, α_0 falls into the range from 0 to 1. For $\alpha_i \leq \alpha_0$, $P_s^{*i}(\alpha_i)$ always takes the value of P_{max}. Hence, U_p reaches the maximum value in that range when α_0 is chosen. For $\alpha_0 < \alpha_i \leq 1$, there exists one and only one root α_r for the above quadratic function, which is in the range from 0 to 1. If $\alpha_r < \alpha_0$, then $\frac{\partial U_p}{\partial \alpha_i} < 0$ when $\alpha_0 < \alpha_i \leq 1$. The derivative of U_p with respect to α_i is monotonically decreasing. Thus, the optimal $\alpha_i^* = \alpha_0$. Otherwise, the optimal $\alpha_i^* = \alpha_r$.

Based on the above analysis, the optimal α_i^* is given by

$$\alpha_i^* = \begin{cases} 1 - \sqrt{1 - \dfrac{2W \left| h_{sb}^i \right|^2}{P_{max} \left| h_{ps}^i \right|^2 c + 2W \left| h_{sb}^i \right|^2 + N_0 c}}, & \text{if } \dfrac{W}{c \ln 2} - \dfrac{N_0}{\left| h_s^i \right|^2} < P_{max} \\[2em] \max \left\{ 2 + \dfrac{2}{\frac{c \ln 2}{W}(P_{max} + \frac{N_0}{\left| h_s^i \right|^2}) - 2}, 1 - \sqrt{1 - \dfrac{2W \left| h_{sb}^i \right|^2}{P_{max} \left| h_{ps}^i \right|^2 c + 2W \left| h_{sb}^i \right|^2 + N_0 c}} \right\}, & \text{otherwise} \end{cases}$$

$$\tag{3.15}$$

This completes the proof.   □

3.3.4 Existence of the Stackelberg Equilibrium

In this section, we prove that the solutions P_s^* in (3.10) and α^* in (3.11) are the Stackelberg Equilibrium. For this purpose, we discuss the two cases with/without considering the power constraint of the SU using the following properties. Since the properties can be easily proved, the detailed proof is omitted here. For both cases, Property 1 always holds as follows, which shows the concavity of the utility function of the SU.

Property 1. The utility function U_s of the SU is concave with respect to its own power level P_s when the time allocation coefficient α is fixed.

We first prove the existence of Stackelberg Equilibrium when the power constraint is not considered. Due to Property 1, U_s is concave with respect to P_s. Without considering the power constraint, setting $\frac{\partial(U_s^i)}{\partial(P_s^i)} = 0$ yields the optimal transmission power P_s^*, which is given in (3.9). With P_s^* in (3.9), the SU can maximize its utility U_s. For the case without considering the power constraint, we also have the following properties.

Property 2. For all SUs, the optimal transmission power P_s^* in (3.9) decreases with the time allocation coefficient α.

Property 3. The utility function of the primary user is concave with respect to the time allocation coefficient α, given that the optimal transmission power P_s^* of the SU in (3.9) is fixed.

Due to Property 2, there is a trade-off for the PU to select the time allocation coefficient α. When the PU allocates less time to the cooperating SU for its own transmission, the SU will choose a lower transmission power during cooperation, which results in a reduction in the utility of the PU. When the PU allocates more time for the SU, the PU will have less time for its own transmission, which may also lead to a decrease in its utility. In other words, the PU cannot keep increasing its utility by increasing α.

Due to Property 3, the optimal α can be obtained by setting $\frac{\partial U_p}{\partial \alpha} = 0$, since the utility function of the PU is concave with respect to α. Therefore, the PU can always find its optimal time allocation coefficient α^* in (3.14) such that $U_p(\alpha^*) \geq U_p(\alpha)$. Together with Property 1, given the time allocation coefficient α, the SU can always find its optimal transmission power P_s^* in (3.9). Then, P_s^* in (3.9) and α^* in (3.14) are the Stackelberg Equilibrium.

In the following, we will discuss the case with power constraint. Due to Property 2, P_s^* in (3.9) increases as α decreases. For a given value of α, P_s^* may achieve its maximum value P_{max}. Since the scenarios before P_s^* approaches P_{max} is the same as the case without power constraint, we only discuss the case when $P_s^* = P_{max}$. When the SU chooses P_{max}, it is optimal for the PU to choose α_0, as in the analysis of α^* in Sect. 3.3.3.2. Therefore, we conclude that the solutions P_s^* in (3.10) and α^* in (3.11) are the Stackelberg Equilibrium.

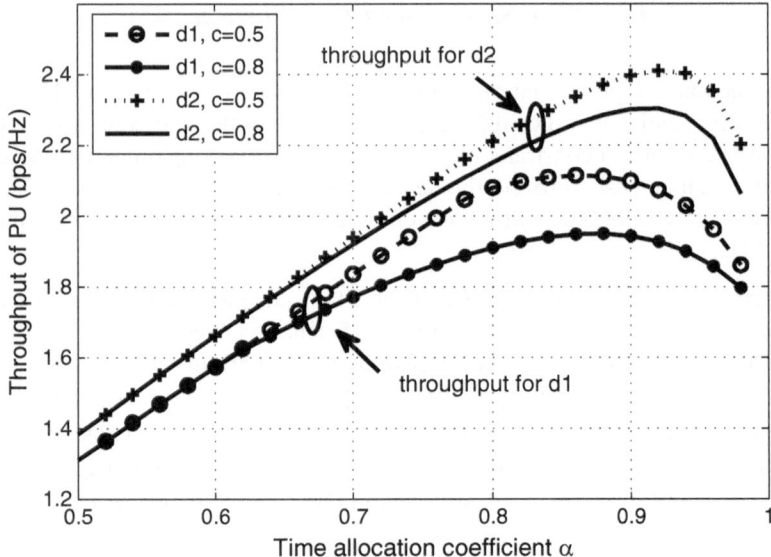

Fig. 3.2 Throughput of the PU versus the time allocation coefficient α

3.3.5   Numerical Results

In this section, we present numerical results so as to provide insight into the proposed cooperation scheme. Similar to [36], normalizing the distance between PU and BS, the SU is approximately placed at the distance $d \in (0, 1)$ from the PU and $1 - d$ from the BS. By doing so, the channel gains can be calculated based on a simple path loss model, e.g., $\left|h_{ps}^i\right|^2 = \frac{1}{d^\zeta}$ and $\left|h_{sb}^i\right|^2 = \frac{1}{(1-d)^\zeta}$, where $\zeta = 3$ is the path loss coefficient. Aiming at reducing the system parameters, the maximum secondary transmission power P_{max} is normalized to 1 and we choose $P_{max}/N_0 = 10$ dB.

Figure 3.2 shows the PU's throughput with respect to α, for different weights and normalized distances d_1 and d_2 ($d_1 > d_2$), respectively. The normalized distance d_s between the SU and its corresponding receiver is set to 0.5. It can be seen that there exists an optimal α^* which can maximize the throughput via cooperation. With a smaller weight, the throughput of the PU is higher. The reason is that with a smaller weight, the SU is more concerned with the throughput than the energy consumption, and hence the SU is willing to spend more power for cooperation. Moreover, it can also be seen that different distances affect different throughput and the optimal α^*. Figure 3.3 shows the optimal parameters α^* versus the normalized distance d, for $c = 0.3$ and 0.6. It is seen that the optimal parameter α^* rises fast as d increases. Further, a greater weight c results in a smaller optimal α^*, which mainly falls in the range from 0.8 to 1.

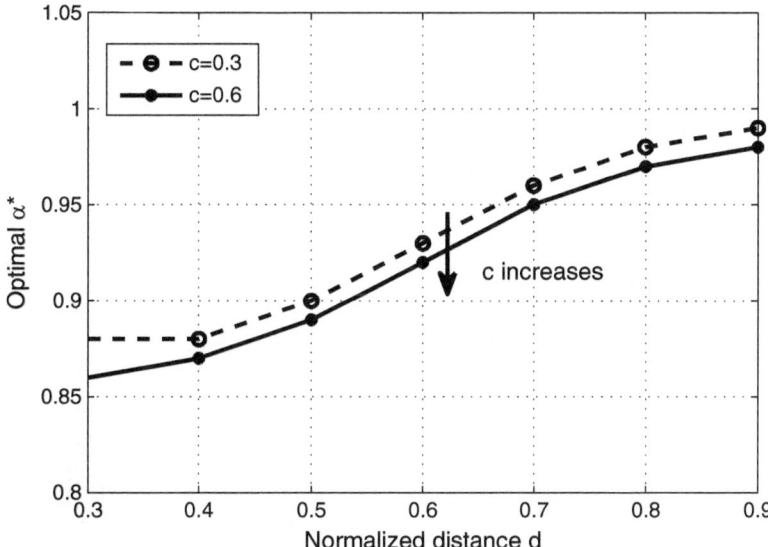

Fig. 3.3 Optimal α^* versus the normalized distance d

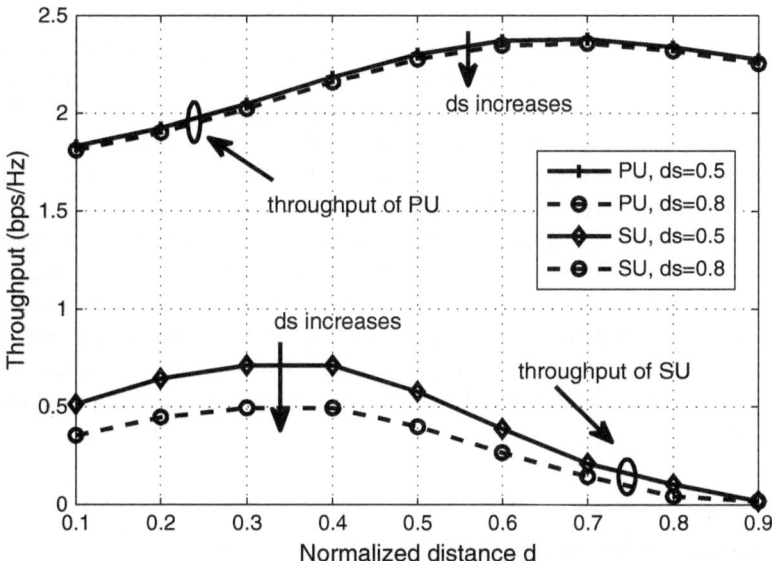

Fig. 3.4 Throughput versus the normalized distance d

Figure 3.4 shows the maximum throughput of the PU and the SU versus the normalized distance d. The throughput is achieved by selecting the optimal α^* and the optimal transmission power $P_s^*(\alpha^*)$. It can be seen that, as d increases, the

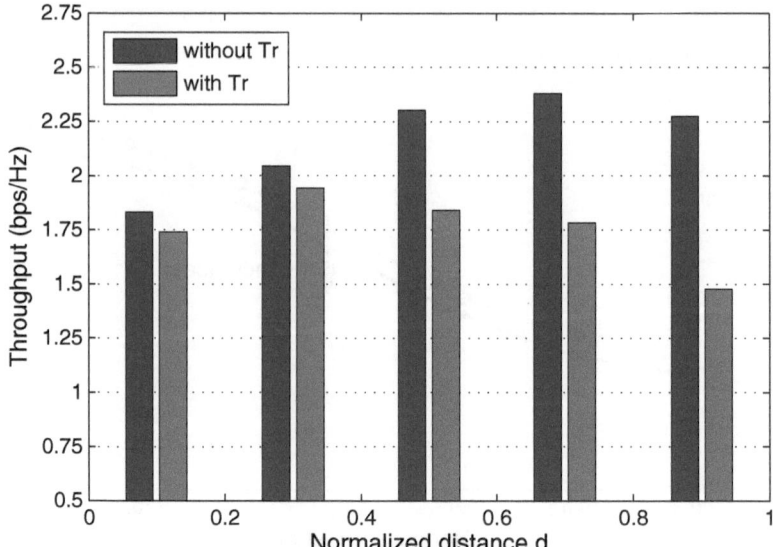

Fig. 3.5 Throughput versus the normalized distance d

maximum throughput of the PU first rises then drops slightly; and the maximum
throughput of the SU drops. This is because the optimal α^* rises almost with d,
the optimal power $P_s^*(\alpha^*)$ decreases accordingly; the throughput of the SU is a
function of $P_s^*(\alpha^*)$, while the throughput of the PU is a function of α^* and $P_s^*(\alpha^*)$.
Furthermore, a greater normalized distance d_s between the SU and its corresponding
receiver leads to a lower throughput for both the PU and the SU.

Figure 3.5 shows the impact of trust values on the SU selection. A number of
SU_i ($i = 1, 2, 3, 4, 5$) with associated trust values 0.95, 0.95, 0.8, 0.75 and 0.65,
are located at the normalized distances $d = 0.1, 0.3, 0.5, 0.7$ and 0.9, respectively.
Without considering trust values, the PU should select SU_4 since the PU can achieve
the highest throughput via cooperation with SU_4. Considering trust values of SUs,
SU_2 is the best choice since the PU can attain the highest expected throughput via
cooperation with SU_2.

3.4   Cooperation for Energy Saving

In this section, we will discuss the cooperation when the primary link is good.
Under this circumstance, the PU can satisfy the QoS requirement (e.g., transmission
rate) on its own, rather than relying on cooperation. However, considering its power
limitation, it might have an incentive to save energy through cooperation, which can
prolong its battery life. We will analyze the cooperation to maximize the energy
saving of the PU when the primary link is good.

3.4.1 Stackelberg Game between PU and SU

We model the cooperation under the same Stackelberg game framework, where the PU acts as leader and the SU acts as follower. Based on the outcomes of Stackelberg game, the PU can choose the best SU and the optimal parameters to maximize the energy saving.

3.4.1.1 Primary User

To meet the required transmission rate R_r, without cooperation, the PU needs to spend P_d for transmission. The following equation holds.

$$R_r = W \log_2 \left(1 + \frac{P_d |h_{pb}|^2}{N_0} \right), \tag{3.16}$$

Suppose that the PU cooperates with SU_i, it selects the time allocation parameter α_i and its transmission power P_c^i to maximize the energy saving. The transmission rate R_c^i through AF cooperative communication between the PU and SU_i is the same as (3.3). In order to satisfy the QoS requirement, the PU needs to maintain the same transmission rate, i.e., R_c^i needs to be equal to R_r.

For the non-cooperative case, the PU uses P_d to transmit for a period of T. Therefore, the energy consumption is $P_d T$ joules. When the PU cooperates with SU_i, it uses P_c^i to transmit for a period of $\frac{\alpha_i T}{2}$. The energy consumption is $P_c^i \frac{\alpha_i T}{2}$. The energy saving for the PU is the ratio between the amount of energy reduction with cooperation and the total power dissipation without cooperation:

$$\gamma_i = \frac{P_d T - P_c^i \frac{\alpha_i T}{2}}{P_d T} = 1 - \frac{\alpha_i P_c^i}{2 P_d}, \tag{3.17}$$

where P_c^i needs to satisfy the power constraint, i.e., $P_c^i \leq P_{max}$.

If γ_i is positive, the PU will choose to cooperate to save energy. Otherwise, the PU will not choose to cooperate since it cannot save energy through cooperation. If $\gamma_i = 0$, it is equivalent to choose cooperation or not. Considering the trust value Tr_i of each neighboring SU_i, from (3.17), the utility function of the PU is given by

$$U_p^i = Tr_i \cdot \left(1 - \frac{\alpha_i P_c^i}{2 P_d} \right), \tag{3.18}$$

which indicates the expected energy saving. The objective of the PU is to maximize its utility function and the strategy is to choose the most suitable SU from the set of its one-hop neighboring SUs and the cooperation parameters, i.e., the time allocation parameters α_i and transmission power P_c^i for cooperation with the selected SU_i.

3.4.1.2   Secondary User

The utility of secondary user is the same as (3.6). The strategy of the SU is to decide its transmission power, given α and P_c, with the target of maximizing transmission rate without expending too much energy.

3.4.2   Game Analysis

To analyze the Stackelberg game, the backward induction method is applied. By doing so, the best response functions of both the PU and the SU are derived. Then, the Stackelberg equilibrium of the proposed game can be achieved based on the best response functions. However, there is no close-form solution for the best response of the PU.

3.4.2.1   Best Response Function of SU

The approach to analyze the best response function of the SU is the same as the one in the previous section. The close-from solution is shown in (3.10).

3.4.2.2   Best Response Function of PU

Based on the best response of the SU, the leader of the Stackelberg game (PU) can optimize its strategy (i, α_i, P_c^i) so as to maximize its utility according to (3.18), being aware that its decision will affect the strategy selected by the Stackelberg follower (SU). The PU decides whom it selects to cooperate with (the parameter i) and how it cooperates with this SU_i (slot allocation parameters α_i and the transmission power P_c^i). According to (3.18), the PU aims at solving the following optimization problem:

$$\max_{\alpha_i, \beta_i, i} \ U_p^i = T r_i \cdot \left(1 - \frac{\alpha_i P_c^i}{2 P_d}\right)$$

$$\text{s.t.} \frac{\alpha_i W}{2} \log_2 \left[1 + \frac{P_c^i |h_{pb}|^2}{N_0} + f\left(P_c^i \left|h_{ps}^i\right|^2, P_s^i \left|h_{sb}^i\right|^2\right)\right] = W \log_2 \left(1 + \frac{P_d |h_{pb}|^2}{N_0}\right)$$

$$0 < \alpha_i \le 1, 0 < P_c^i \le P_{max}, SU_i \subseteq \mathbf{S}_p,$$

where the set $\mathbf{S}_p$ is the collection of the one-hop neighboring SUs of the PU. Let $(i^*, \alpha_{i^*}^*, P_c^{*i^*})$ be the solution of this optimization problem. Thus, the optimal strategy is given by

$$i^*, \alpha_{i^*}^*, P_c^{*i^*} = \arg \max_{\alpha_i, P_c^{*i^*}, i} U_p^i.$$

However, there is no close-form solution for this non-linear optimization problem.

Using the backward induction method, the PU can determine the optimal strategies α_i^*, P_c^{i*} when cooperation with SU_i. The best SU_{i*} will be selected so that PU will have the highest expected energy saving when it cooperates with this SU. For the selected SU_{i*}, we observe that it has the unique best response $P_s^{i*}(\alpha_{i*}^*)$ to the cooperation parameters the PU chooses and the PU has a unique best choice $(\alpha_{i*}^*, P_c^{*i^*})$. Thus, for SU_{i*} and the PU, the Stackelberg equilibrium of the game is $(\alpha_{i*}^*, P_c^{*i^*}, P_s^{i*}(\alpha_{i*}^*))$. The PU's best strategy is to choose $(\alpha_{i*}^*, P_c^{*i^*}, i^*)$ and SU_{i*}'s best strategy is to choose $P_s^{i*}(\alpha_i^*)$ as its optimal transmission power.

Algorithm 1 represents the detailed procedure for cooperation under different channel conditions.

Algorithm 1 SU Selection Algorithm

1: **if** Channel is good **then**
2:    **for** each $SU_i \subseteq \mathbf{S}_p$ **do**
3:       calculate the transmission power P_s^i of SU_i;
4:       calculate the cooperative transmission rate R_c^i;
5:       calculate α_i^*, P^{*i} such that the energy saving of PU achieves the optimal γ_i^*;
6:       $U_p^i = Tr_i \cdot \gamma_i^*$;
7:    **end for**
8:    **for** each $SU_i \subseteq \mathbf{S}_p$ **do**
9:       find i^* such that $U_p^{i^*} = \max U_p^i$;
10:    **end for**
11:    **if** $U_p^i < 0$ **then**
12:       PU does not choose cooperative mode;
13:    **else**
14:       PU selects SU_i and α_{i*}^*, $P_c^{*i^*}$ for cooperation.
15:    **end if**
16: **else**
17:    **for** each $SU_i \subseteq \mathbf{S}_p$ **do**
18:       calculate the transmission power P_s^i of SU_i;
19:       calculate α_i^* such that the throughput of the PU achieves its optimal value R_c^{i*};
20:       $U_p^i = Tr_i \cdot R_c^{i*}$;
21:    **end for**
22:    **for** each $SU_i \subseteq \mathbf{S}_p$ **do**
23:       find i^* such that $U_p^{i^*} = \max U_p^i$;
24:    **end for**
25:    **if** $U_p^{i^*} < R_r$ **then**
26:       PU does not choose cooperative mode;
27:    **else**
28:       PU selects SU_i and α_{i*}^*, $P_c^{*i^*}$ for cooperation.
29:    **end if**
30: **end if**

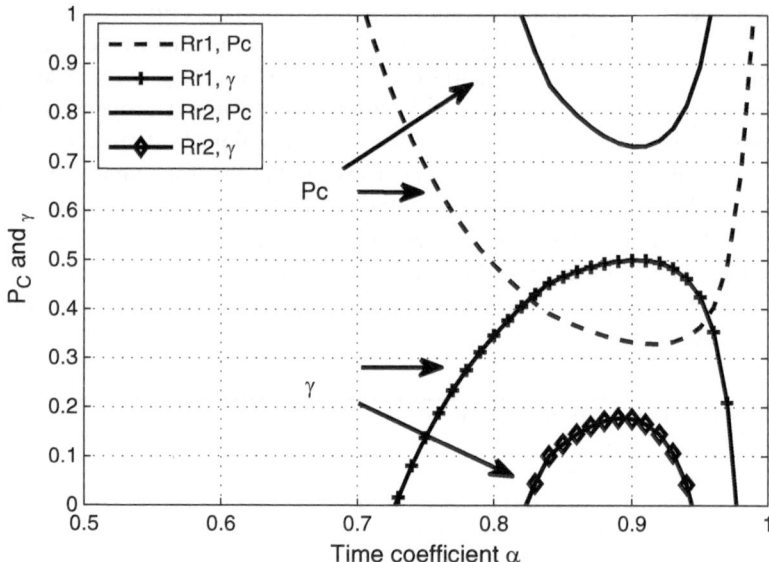

Fig. 3.6 P_C, γ versus the time allocation coefficient α

3.4.3   Numerical Results

In this section, we present numerical results to evaluate the performance of the proposed cooperation scheme for energy saving. Similar to the previous section, normalizing the distance between PU and BS, the SU is approximately placed at the distance $d \in (0, 1)$ from the PU and $1 - d$ from the BS. By doing so, the channel gains can be calculated based on a simple path loss model, e.g., $\left|h^i_{ps}\right|^2 = \frac{1}{d^\zeta}$ and $\left|h^i_{sb}\right|^2 = \frac{1}{(1-d)^\zeta}$, where $\zeta = 3$ is the path loss coefficient. Aiming at reducing the system parameters, the maximum secondary transmission power P_{max} is normalized to 1 and we choose $P_{max}/N_0 = 10$ dB.

Figure 3.6 shows the trends of energy saving γ and transmission power P_c with respect to α for different rate requirements ($R_1 < R_2$). The normalized distance $d = 0.5$ and the weight $c = 0.4$. The solid line and dashed line represent γ and P_c, respectively. The PU chooses cooperation only when it can save energy, i.e., γ should be positive. The PU has the power constraint, i.e., $P_c \leq 1$. It can be seen that there exists an optimal α that the PU can select to achieve the maximum energy saving. It can also be seen that a smaller required transmission rate results in a higher energy saving and wider cooperation range in terms of α, in which the PU can save energy by selecting α in that range.

Figure 3.7 shows the trends of parameters α^*, P_c^* and γ^* versus the normalized distance d. It can be seen that the optimal value P_c^* and α^* tend to increase with distance d. The reason is that the channel gain decreases with increasing distance between the PU and the SU so that the PU requires a larger power for primary

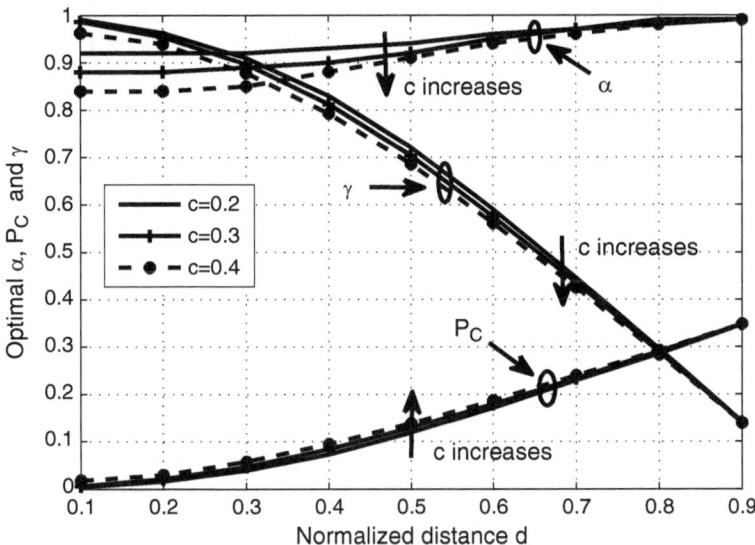

Fig. 3.7 α^*, P_c^*, γ^* versus the normalized distance d

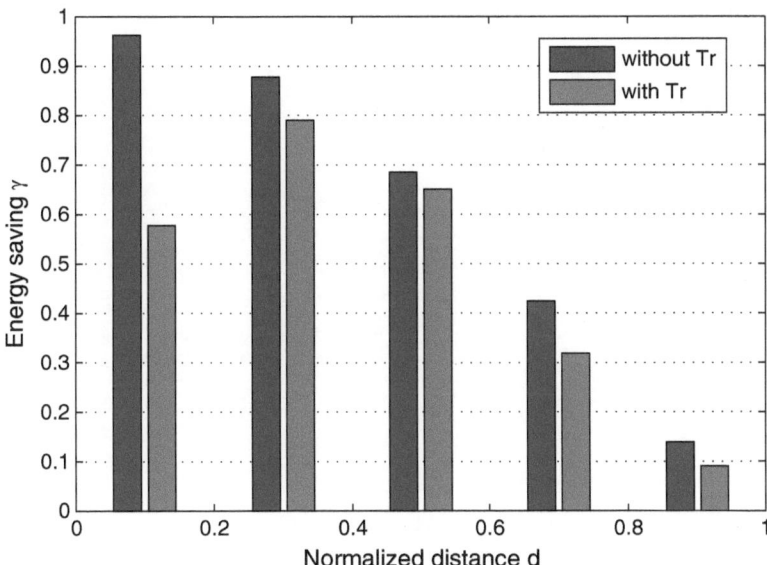

Fig. 3.8 Energy saving versus the normalized distance d

transmission. Moreover, with increasing d, the benefit that the PU can obtain from the cooperation (the energy saving) becomes less, because the PU needs to spend more (P_c^*) power and time (α^*) for cooperation.

Figure 3.8 shows the impact of trust values on the SU selection. A number of SU_i ($i = 1, 2, 3, 4, 5$) with associated trust values 0.6, 0.90, 0.95, 0.75 and 0.65, are located at the normalized distances $d = 0.1, 0.3, 0.5, 0.7$, and 0.9, respectively. Without considering trust values, the PU should select SU_1 since the PU can achieve the highest energy saving via cooperation with SU_1. However, considering trust values of SUs, SU_2 is the best choice since the PU can attain the highest expected energy saving.

3.5   Summary

In this chapter, we have proposed a novel cooperation scheme for CRNs in an unfriendly environment. The PU chooses the best SU and the optimal cooperation parameters to maximize its throughput or energy efficiency, based on the channel condition, considering the trustworthiness of SUs. While the selected SU determines the optimal power for cooperative and secondary transmissions. We formulate the procedure of decision making as a Stackelberg game. The analysis of the game provides the PU with the best strategy for the SU selection, spectrum access time allocation and transmission power determination. Numerical results demonstrate that, with the proposed scheme, by selecting the most suitable SU, the PU can attain higher throughput or save energy through cooperation.

Chapter 4
Cooperative Networking for Secure Communications

Abstract In this chapter, we investigate the cooperation schemes in cognitive radio networks, which target to help the primary users for secure communication to deal with the issues of eavesdropping and provide transmission opportunities to secondary users. Two types of cooperation schemes are proposed with the objective of maximizing the secrecy rate of the primary user, and different scenarios in terms of the availability of information regarding eavesdroppers are considered. Simulation results are given to validate the proposed schemes.

4.1 Introduction

In the previous chapter, we aim at improving the performance of cooperation in the presence of untrusted SUs. In addition, in such an insecure environment, there may exist some unfriendly users, e.g., eavesdroppers. Due to the broadcast nature of wireless communication, these unfriendly users can easily overhear the ongoing transmission. Since security is a critical issue in wireless environments due to the broadcast nature of wireless communications [88], PUs also have the need for secure communications. Traditionally, the security is dealt with by encryption at upper layers; yet, it becomes very challenging for a network without infrastructure [74]. Moreover, the encryption algorithms could be compromised and an alternative way for enhancing the security is to protect the transmitted signal from being received or decoded by the eavesdroppers [89]. Recently, physical (PHY) layer security, or information-theoretic security, has attracted a lot of attentions in the research community [75, 90, 91], which exploits the properties of the wireless channel to secure communications. In [75], it is shown that the perfectly secure information can be transmitted at a nonzero rate from the source to the destination, while the eavesdropper cannot learn anything regarding it. This rate is referred to the *secrecy rate*, which is defined as the difference between the transmission rate of the source-destination link and that of the source-eavesdropper link. However, the secrecy rate would be equal to zero when the source-destination channel is worse than the source-eavesdropper channel.

N. Zhang and J.W. Mark, *Security-aware Cooperation in Cognitive Radio Networks*,
SpringerBriefs in Computer Science, DOI 10.1007/978-1-4939-0413-6_4,
© The Author(s) 2014

To address the above issue, user cooperation has been introduced to enhance the secrecy of communications [76, 78, 92–94]. In [78], three types of schemes using decode-and-forward (DF), amplify-and-forward (AF), and cooperative jamming, are proposed to improve the secrecy via cooperation. In [92], distributed beamforming is leveraged at relays to enhance the source's secrecy. Nevertheless, these schemes cannot be applied directly to CRNs because the special features of CRNs have not been taken into consideration: (1) PUs have higher priorities for spectrum usage in CRNs; (2) it might not be reasonable to assume that PUs and SUs cooperate unconditionally with each other, since they have their own interests. Considering the features of CRNs, a cooperation based spectrum access is studied in [70], which improves the security of the primary link and provides transmission opportunities to SUs. However, the cooperation objective is achieved at the expense of employing multiple antennas and only the scenario with a single eavesdropper is considered. In reality, the assumption of multiple antennas might not be feasible. Moreover, more practical scenarios, where there exist multiple eavesdroppers or the information regarding eavesdropper(s) is not available, need to be investigated.

In this chapter, we will investigate cooperation schemes for CRNs, whereby the SUs can cooperate with the PU to enhance the security of the primary link and in return access the channel as a reward for their own transmission. The cooperation is performed on a mutual benefit basis, where both parties can benefit from their cooperation. The scenario under consideration includes an ad hoc primary network and an ad hoc secondary network, with some eavesdroppers in the network. To enhance the security of primary link, the PU can either cooperate with two individual SUs (as a relay and a jammer), or a cluster of SUs, which are referred to as relay-jammer (R-J) scheme and cluster-beamforming (C-B) scheme, respectively. For the former, the relay SU employs decode-and-forward (DF) mode to transmit the PU's information, and in the meanwhile the jammer SU creates artificial noise to confound the eavesdropper. To achieve the maximum secrecy rate,[1] joint time and power allocation is studied. For the latter, the SUs in the cluster enhance the secrecy of the PU's communication via collaborative beamforming. Three different approaches are proposed for the scenarios with one eavesdropper, with multiple eavesdroppers, and without eavesdroppers' information, respectively. With the objective of maximizing the secrecy rate, the weight selection and time allocation are investigated. Moreover, two different scenarios with one eavesdropper and with multiple eavesdroppers are discussed. Unlike previous work on cooperation for physical layer security, which lacks incentive for cooperation, this work is based on mutual benefits to motivate all the participants for cooperation. Through cooperation, the PU can enhance the security of its communication and the SUs can earn transmission opportunities, which creates a win-win situation.

---

[1] Secrecy rate is defined as the difference between the transmission rate of the source-destination link and that of the source-eavesdropper link.

4.2 System Model

Shown in Fig. 4.1, the system consists of a primary source (S), a primary destination (D), multiple SUs, and one eavesdropper (E) or multiple eavesdroppers who intends to decode the PU's information [95]. In the primary network, S holds a time slot with duration T to communicate with D over a bandwidth of W Hz. Different from [95, 96], which assume that there is no direct link between S and either D or E, and only focus on the secure information transfer from the relays to D, a more general case is considered, where direct links from S to D and from S to E exist. When the channel between S and D is worse than that between S and E, the secrecy rate is zero. In order to transfer information securely, S either chooses two cooperating SUs, i.e., a relay SU (R) and a jammer SU (J), or a cluster (C) of SUs for cooperation, which are all considered trustworthy [78].

Cooperation can be performed in a three-phase fashion or a two-phase fashion. Fig 4.1a shows the time structure for the three-phase cooperation. A fraction α of the duration T is used for the transmission from S to D, which is further divided into two parts according to β, where $0 < \alpha, \beta < 1$. Particularly, in the first phase with duration of $\alpha(1-\beta)T$, S transmits data to cooperating SUs, which is also overheard by D and E because there exist direct links. In the second phase, a subsequent duration of $\alpha\beta T$ is leveraged for the transmission from cooperating SUs to D. For R-J cooperation scheme (Fig. 4.1a), R employs DF protocol to relay the PU's message to D, and simultaneously J transmits an artificial jamming signal which is independent of the PU's message to confound E. For C-B cooperation scheme (Fig. 4.1b), the SUs in C first decode the PU's message and then each of them forwards a weighted version of that message to D via collaborative beamforming. In the last phase, the remaining $(1-\alpha)T$ is granted to cooperating SUs for transmitting their own data to the corresponding secondary receivers as a reward. The relay SU and jammer SU access the channel in a TDMA fashion, while the SUs in C transmit the data to a common secondary receiver via collaborative beamforming [97]. When there exist multiple eavesdroppers, C-B cooperation scheme is carried out in a two-phase fashion, as shown in Fig. 4.1b. The operation in the first phase is the same as

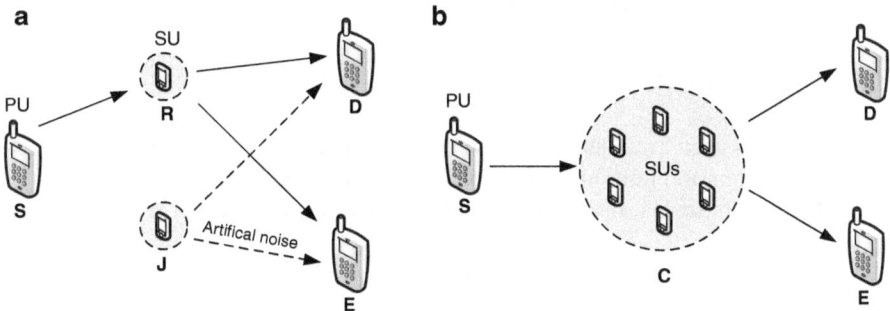

Fig. 4.1 The system model. (**a**) R-J cooperation scheme, (**b**) C-B cooperation scheme

previous cases. In the second phase, the cluster simultaneously transmits the PU's message and its own data. Note that, in this paper, R-J cooperation scheme is only considered for the scenario where there exists only one eavesdropper.

In R-J cooperation scheme, R uses DF strategy to increase the transmission rate at D and J increases the interference at E. To maximize the secrecy rate, the PU and cooperating SUs jointly optimize the time and transmission power allocation. Specifically, S determines the time allocation coefficients α and β, while R and J choose the transmission power for cooperation. In C-B cooperation scheme, the PU and cooperating SUs jointly allocate the time and beamforming weights to maximize the secrecy rate, by considering that the SUs in C as a whole have a total power constraint.

A slow, flat, block Rayleigh fading environment is considered, where the channel remains static in one time slot and changes independently over different slots. The channel coefficient from S to D is denoted by h_{SD}. Similarly, we have h_{SR}, h_{SE}, h_{RD}, h_{RE}, h_{JD}, and h_{JE}. The global CSI is assumed available in the system, including D-related CSI (D-CSI) and E-related CSI (E-CSI), which is a common practice in the physical layer security literature. The cooperation when E-CSI is unavailable will be discussed in Sect. 4.5. In addition, additive white Gaussian noise is assumed with zero mean and one-side power spectrum density N_0. Moreover, each node is equipped with a single antenna and communicates with each other in a half-duplex mode.

In the following, matrices and vectors are denoted by bold uppercase letters and bold lowercase letters, respectively. $(\cdot)^*$, $(\cdot)^T$, and $(\cdot)^\dagger$ denote the conjugate, transpose, and conjugate transpose, respectively. $\mathbf{I}$ denotes the identity matrix. $[x]^+$ denotes the maximum value between x and 0, while $x^\star$ denotes the optimal value of x. $|\cdot|$ denotes the magnitude of a channel or the absolute value of a complex number, while $\|\cdot\|$ is the Euclidean norm of a vector or a matrix. The key notation can be found in Table 4.1.

4.3 R-J Cooperation Scheme

4.3.1 Problem Formulation

4.3.1.1 Secrecy Rate of PU

Secrecy rate is used as a measure for secure communication. To obtain the secrecy rate, the transmission rates at different nodes are calculated as follows.

In the first phase, S transmits data to R and the transmission rate at R is given by

$$R_R = W \log_2(1 + \gamma), \tag{4.1}$$

where $\gamma = \frac{P|h_{SR}|^2}{WN_0}$ and P is the transmission power of the PU.

Table 4.1   The main notations

Symbol	Description
S	The primary source
D	The primary destination
R	The relay SU
J	The jammer SU
E	Eavesdropper
C	The cluster of SUs
E-CSI	The channel state information (CSI) regarding E
$\bar{R}_{EX}$	The expected overall transmission rate of SUs
P_{max}^{C}	The maximum power of SU(s) for cooperation
P_{max}	The power budget of individual user or the cluster
$\bar{R}_{S,i}$	The overall transmission rate of SU_i
$U_{S,i}$	The satisfaction of SU_i for the transmission rate via cooperation
h_{SD}	The channel gain from S to D
h_{SE}	The channel gain from S to E
h_{SR}	The channel gain from S to R
h_{RD}	The channel gain from R to D
h_{RE}	The channel gain from R to E
h_{JD}	The channel gain from J to D
h_{JE}	The channel gain from J to E
α, β	The access time allocation coefficient
$\bar{R}_Q$	The predefined required transmission rate of the PU for the scenario without E-CSI
P_R	The transmission power of R during cooperation
P_J	The transmission power of J during cooperation
$\bar{R}_{SEC}$	The overall secrecy rate of the PU

In the second phase, R relays the PU's message to D using DF protocol, and simultaneously J broadcasts an artificial jamming signal. Since D receives signals in both the first and second phases, the transmission rate R_D at D using maximal ratio combining (MRC) is given by

$$R_D = W \log_2 \left(1 + \xi + \frac{P_R |h_{RD}|^2}{W N_0 + P_J |h_{JD}|^2} \right), \tag{4.2}$$

where $\xi = P|h_{SD}|^2/(W N_0)$ is the SNR from the first phase, and P_R and P_J are the transmission power of R and J during cooperation, respectively.

Similarly, E also receives signals during the first two phases. Therefore, the transmission rate at E can be expressed as follows:

$$R_E = W \log_2 \left(1 + \delta + \frac{P_R |h_{RE}|^2}{W N_0 + P_J |h_{JE}|^2} \right), \tag{4.3}$$

where $\delta = P|h_{SE}|^2/(W N_0)$ is the SNR from the first phase.

When DF cooperative communication is applied, the overall transmission rate of D and E equals the minimum rate of the first two phases [98], i.e.,

$$\bar{R}_D = \min\{\alpha(1-\beta)R_R, \alpha\beta R_D\} \tag{4.4}$$

$$\bar{R}_E = \min\{\alpha(1-\beta)R_R, \alpha\beta R_E\} \tag{4.5}$$

As per the definition, the *secrecy rate* R_{SEC} is given by:

$$R_{SEC} = [R_D - R_E]^+, \tag{4.6}$$

Substituting (4.5) into (4.6), the overall secrecy rate is then given by

$$\bar{R}_{SEC} = [\min\{\alpha(1-\beta)R_R, \alpha\beta R_D\} - \alpha\beta R_E]^+ \tag{4.7}$$

4.3.1.2 Overall Transmission Rate of SUs

Let $P_{S,i}$ be the transmission power of SU_i for its own communication, where $i = R$ or J. SUs are considered to have the same power constraint P_{max}. R and J transmit in a TDMA mode and the overall transmission rate of SU_i is given by

$$\bar{R}_{S,i} = \frac{1-\alpha}{2} W \log_2\left(1 + \frac{P_{S,i}|h_{S,i}|^2}{W N_0}\right), \tag{4.8}$$

where $h_{S,i}$ is the channel coefficient from i to its corresponding receiver. Note that each SU has an expected overall transmission rate $\bar{R}_{EX}$, which is the rate that the SU desires through cooperation. However, $\bar{R}_{S,i}$ depends on the time period granted by the PU. From the PU's perspective, it tends to grant less time to SUs, and hence $\bar{R}_{S,i}$ may be much less than $\bar{R}_{EX}$. To measure SUs' degree of satisfaction, $U_{S,i}$ is defined as $U_{S,i} = \min\{\frac{\bar{R}_{S,i}}{\bar{R}_{EX}}, 1\}$, which implies how satisfactory SU_i is with $\bar{R}_{S,i}$. For instance, if $\bar{R}_{S,i} = \bar{R}_{EX}$, $U_{S,i}$ is equal to 1.

In order to enforce the PU to grant an acceptable rewarding time, the SU determines the efforts that it is willing to make during cooperation, i.e., the maximum power P_{max}^C for cooperation. For simplicity, $P_{max}^C = U_{S,i} \cdot P_{max}$. In other words, the degree of efforts that the SU is willing to make depends on the degree of the satisfaction obtained. For example, if $U_{S,i} = 1$, the SU is willing to devote full power P_{max} for cooperation, i.e., $P_{max}^C = P_{max}$.

4.3.1.3 Secrecy Rate Maximization

Since the SU typically does not have much transmission opportunities, it aims at maximizing the throughput by adopting P_{max} for its own transmission. Thus, given a certain α, $\bar{R}_{S,i} = \frac{1-\alpha}{2} W \log_2(1 + \frac{P_{max}|h_{S,i}|^2}{W N_0})$. Based on the degree of the satisfaction,

P_{max}^C can be determined, which is a function of α. As shown in (4.7), $\bar{R}_{SEC}$ is related to α, β, and the transmission power P_R and P_J, which are constrained by P_{max}^C. From PU's perspective, the objective of cooperation is to maximize the overall secrecy rate $\bar{R}_{SEC}$. Therefore, the PU chooses the time allocation coefficients α and β, while the SUs determine the optimal transmission power for cooperation, which can be formulated as the following optimization problem:

$$\max_{\alpha, \beta, P_R, P_J} \bar{R}_{SEC} \tag{4.9}$$

$$s.t.\ 0 < \alpha,\ \beta < 1,\ 0 \le P_R \le P_{max}^C,\ 0 \le P_J \le P_{max}^C. \tag{4.10}$$

4.3.2   Cooperation Parameters Determination

The time allocation coefficients and transmission power can be optimized by solving the above optimization problem. To do this, the procedure can be divided into two steps: (1) given α, R and J select the optimal transmission power; and (2) S selects the optimal $\alpha^\star$, $\beta^\star$ to maximize the secrecy rate, aware of the results of the first step.

From (4.7), for a given α, the overall secrecy rate $\bar{R}_{SEC}$ not only depends on $R_D - R_E$, but also on β. In fact, $\bar{R}_{SEC}$ can be further expressed as follows:

$$\bar{R}_{SEC} = [\alpha\beta(R_D - R_E)]^+ = \alpha\left[\frac{R_R(R_D - R_E)}{R_R + R_D}\right]^+ = \alpha\left[R_R - \frac{R_R(R_R + R_E)}{R_R + R_D}\right]^+,$$

$$\tag{4.11}$$

where R_R, R_D and R_E are given by (4.1), (4.2), and (4.3), respectively. The derivation is given in the following proof. Note that given α, the optimal $\beta^\star = \frac{R_R}{R_R + R_D}$.

Proof. When $\alpha(1 - \beta)R_R \ge \alpha\beta R_D$, we have $\beta \le \frac{R_R}{R_R + R_D}$. Then, the secrecy rate in (4.7) can be given by $[\alpha\beta R_D - \alpha\beta R_E]^+ = \alpha\beta[(R_D - R_E)]^+$, which is a monotonically increasing function with respect to β. To maximize the secrecy rate, β should take the maximum value $\frac{R_R}{R_R + R_D}$. Substituting $\beta = \frac{R_R}{R_R + R_D}$ into (4.7), the secrecy rate can be rewritten as follows: $\bar{R}_{SEC} = \alpha[\frac{R_R(R_D - R_E)}{R_R + R_D}]^+$. When $\alpha(1 - \beta)R_R \le \alpha\beta R_D$, we have $\beta \ge \frac{R_R}{R_R + R_D}$. Then, the secrecy rate in (4.7) can be given by $[\alpha(1 - \beta)R_R - \alpha\beta R_E]^+$. which is a monotonically decreasing function of β. To maximize the secrecy rate, β should take the minimum value $\frac{R_R}{R_R + R_D}$. Substituting $\beta = \frac{R_R}{R_R + R_D}$ into (4.7), the secrecy rate can be rewritten as follows: $\bar{R}_{SEC} = \alpha[\frac{R_R(R_D - R_E)}{R_R + R_D}]^+$. As shown above, for the two cases, to maximize the $\bar{R}_{SEC}$, β always equals to $\frac{R_R}{R_R + R_D}$. Moreover, when β takes the optimal value, it holds that $\alpha(1 - \beta)R_R = \alpha\beta R_D$. Thus, $\bar{R}_{SEC} = \alpha[\frac{R_R(R_D - R_E)}{R_R + R_D}]^+ = \alpha[R_R - \frac{R_R(R_R + R_E)}{R_R + R_D}]^+$.

This completes the proof. $\square$

In the literature, most of the existing works assume the time durations for the transmission from S to R and from R to D are equal, and try to maximize $R_D - R_E$ based on this assumption. However, R_R and R_D are typically not the same. Furthermore, the overall transmission rate is the minimum one between $\bar{R}_R$ and $\bar{R}_D$ for DF strategy. Thus, it is not optimal to assign equal duration for these two phases. From (4.11), it can be seen that the secrecy rate cannot achieve the optimum value by only maximizing $R_D - R_E$. This is because when R_D increases, $R_D - R_E$ increases, but β decreases. Note that the objective function in (4.11) has considered the above factors and in this chapter we study the nontrivial case where the secrecy rate is positive.

4.3.2.1 Power Allocation

Since the relay is leveraged to increase the transmission rate at destination compared with that at the eavesdropper, it requires that $|h_{RD}| > |h_{RE}|$. The job of the jammer is to create more interference at the eavesdropper than at the destination and it is necessary that $|h_{JE}| > |h_{JD}|$. In what follows, to achieve the maximum secrecy rate, the optimal transmission power of relay SU and jammer SU are analyzed, when α is given.

Relay SU

Since R_R is fixed, maximizing $\bar{R}_{SEC} = R_R - R_R(R_R + R_E)/(R_R + R_D)$ is equivalent to minimizing $f(P_R, P_J) = (R_R + R_E)/(R_R + R_D)$. Similar to [77], we study the case in the low SNR regime, where we approximate $\log_2(1 + snr) \approx snr$ [99]. Based on (4.1), (4.2), (4.3), and the approximation, we have

$$f(P_R, P_J) = \frac{\Psi_E + P_R|h_{RE}|^2/(WN_0 + P_J|h_{JE}|^2)}{\Psi_D + P_R|h_{RD}|^2/(WN_0 + P_J|h_{JD}|^2)}, \tag{4.12}$$

where $\Psi_D = \gamma + \xi$ and $\Psi_E = \gamma + \delta$. Take the first order derivative of f with respect to P_R, which is always negative because $|h_{RD}| > |h_{RE}|$. Therefore, $f(P_R, P_J)$ is a monotonically decreasing function of P_R and the optimal transmission power $P_R^\star$ is P_{max}^C for maximizing the secrecy rate. Note that P_{max}^C is a function of α.

Jammer SU

The optimal transmission power $P_J^\star$ is selected such that the objective function in (4.12) can be maximized. The derivative of (4.12) with respect to P_J is proportional to a quadratic function in the following form:

$$\frac{\partial f}{\partial P_J} \propto \psi_1 \cdot P_J^2 + \psi_2 \cdot P_J + \psi_3, \tag{4.13}$$

where

$$\psi_1 = |h_{JD}||h_{JE}|P_R(|h_{RD}|\Psi_E|h_{JE}| - |h_{RE}|\Psi_D|h_{JD}|)$$

$$\psi_2 = 2|h_{JD}||h_{JE}|N_0P_R(|h_{RD}|\Psi_E - |h_{RE}|\Psi_D)$$

$$\psi_3 = |h_{RD}||h_{RE}|P_R^2N_0(|h_{JD}| - |h_{JE}|) + N_0^2(|h_{RD}|\Psi_E|h_{JE}| - |h_{RE}|\Psi_D|h_{JE}|).$$

Since $|h_{RD}| > |h_{RE}|$ and $|h_{JE}| > |h_{JD}|$, we have $\psi_1 > 0$, $\psi_2 > 0$, and $P_R = P_{max}^C$. If $\psi_3 > 0$, there is no positive root for the quadratic function in (4.13) and $\frac{\partial f}{\partial P_J} > 0$ for the range from 0 to P_{max}^C. Thus, $P_J^\star$ equals to 0 to maximize the secrecy rate, indicating a non-jamming scenario. If $\psi_3 < 0$, there is one positive root $\frac{-\psi_2+\sqrt{\psi_2^2-4\psi_1\psi_3}}{2\psi_1}$. When $\frac{-\psi_2+\sqrt{\psi_2^2-4\psi_1\psi_3}}{2\psi_1} > P_{max}^C$, $\frac{\partial f}{\partial P_J} < 0$ for the range from 0 to P_{max}^C and hence $P_J^\star$ should be selected as P_{max}^C. Otherwise, $P_J^\star$ should be equal to $\frac{-\psi_2+\sqrt{\psi_2^2-4\psi_1\psi_3}}{2\psi_1}$.

4.3.2.2 Time Allocation

From (4.11), the objective function has taken the factor β into consideration. Given α, the optimal transmission power of SUs has been obtained in the previous section. Therefore, the optimal $\beta^\star$ can be easily determined by

$$\beta^\star = \frac{R_R}{R_R + R_D}, \tag{4.14}$$

where R_D is the transmission rate at D when R and J choose the optimal transmission power.

The optimal $\alpha^\star$ can be determined by solving the following equation:

$$\alpha^\star = \arg\max \alpha\beta(R_D - R_E) \tag{4.15}$$

Note that β, R_D, and R_E are all functions of α ($0 < \alpha < 1$).

4.4 C-B Cooperation Scheme with E-CSI

In this section, we discuss the cooperation between the PU and a cluster of SUs when E-CSI is available. We propose a three-phase cooperation scheme and a two-phase cooperation scheme for the scenarios in the presence of an eavesdropper and multiple eavesdroppers, respectively. To maximize the secrecy rate, time allocation and weights selection are jointly considered.

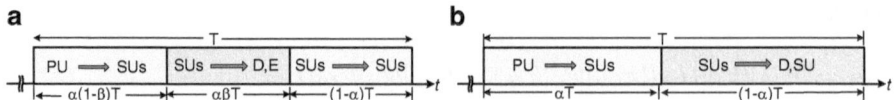

Fig. 4.2 Time frame structure for cooperation. (**a**) Three-phase cooperation (**b**) Two-phase cooperation

4.4.1 C-B Scheme for Single Eavesdropper (CBSE)

4.4.1.1 Problem Formulation

Secrecy Rate of PU

In the presence of one eavesdropper, the cooperation is performed in a three-phase fashion, as shown in Fig. 4.2a. In the first phase, the PU broadcasts to the cluster the signal $\sqrt{P}s$, where s is the information symbol with $E\{|s|^2\} = 1$, which is overheard by D and E. In order for all the cluster members to successfully decode the signal, the transmission rate R_R from S to C is determined by the worst channel between S and the cluster members.

$$R_R = W \log_2 \left(1 + \min_i \frac{P|h_{SR,i}|^2}{N_0 W}\right), \tag{4.16}$$

where $h_{SR,i}$ is the channel from S to ith SU in the cluster. Denote by $y_{D,1}$ and $y_{E,1}$ the signal received at D and E in the first phase, respectively, which can be given by

$$y_{D,1} = \sqrt{P}h_{SD}s + n_{SD} \tag{4.17}$$

$$y_{E,1} = \sqrt{P}h_{SE}s + n_{SE} \tag{4.18}$$

where n_{SD} and n_{SE} are the noise at D and E, respectively.

In the second phase, each SU in the cluster decodes the received symbol and forwards a weighted version of the re-encoded symbol $\tilde{s}$ to D. Let **w** be the column vector of the weights of all SUs in the cluster and N be the number of SUs in the cluster. Then, the received signals $y_{D,2}$ and $y_{E,2}$ at D and E in the second phase can be written respectively as:

$$y_{D,2} = \mathbf{h}_{RD}^\dagger \mathbf{w}\tilde{s} + n_{RD} \tag{4.19}$$

$$y_{E,2} = \mathbf{h}_{RE}^\dagger \mathbf{w}\tilde{s} + n_{RE} \tag{4.20}$$

where $\mathbf{h}_{RD} = [h_{D,1}^*, h_{D,2}^*, \ldots, h_{D,N}^*]^T$ and $\mathbf{h}_{RE} = [h_{E,1}^*, h_{E,2}^*, \ldots, h_{E,N}^*]^T$. Note that $h_{D,i}$ and $h_{E,i}$ are the complex channel coefficients from the ith SU in the cluster to D and E, respectively, where $i \in \{1, 2, \ldots, N\}$. n_{RD} and n_{RE} are the noise at D and E, respectively.

Assume that the cooperating SUs use the same codewords as S. The transmission rate at D and E are given as follows:

$$R_D = W \log_2 \left(1 + \xi + \frac{\mathbf{w}^\dagger \mathbf{h}_{RD} \mathbf{h}_{RD}^\dagger \mathbf{w}}{N_0 W}\right) \qquad (4.21)$$

$$R_E = W \log_2 \left(1 + \delta + \frac{\mathbf{w}^\dagger \mathbf{h}_{RE} \mathbf{h}_{RE}^\dagger \mathbf{w}}{N_0 W}\right), \qquad (4.22)$$

where ξ and δ are the same as that in (4.2) and (4.3), respectively. Substituting (4.16) and (4.22) into (4.7), we can obtain the overall secrecy rate.

Overall Transmission Rate of SUs

In the third phase, the SUs in the cluster transmit the data to the secondary receiver via collaborative beamforming. The overall rate $\bar{R}_S$ at the secondary receiver can be given by

$$\bar{R}_S = (1 - \alpha) W \log_2(1 + \frac{\mathbf{v}^\dagger \mathbf{h}_{RS} \mathbf{h}_{RS}^\dagger \mathbf{v}}{N_0 W}), \qquad (4.23)$$

where $\mathbf{v}$ is the column vector of the weights of all cooperating SUs for the secondary transmission and $\mathbf{h}_{RS} = [h_{S,1}^*, h_{S,2}^*, \ldots, h_{S,N}^*]^T$. Note that $h_{S,i}$ is the complex channel coefficient from the ith SU in the cluster to the secondary receiver. To maximize the transmission rate, the SUs select the optimal $\mathbf{v}^\star$, under the total power constraint, which can be formulated as follows:

$$\max_{\mathbf{v}} \ \mathbf{v}^\dagger \mathbf{h}_{RS} \mathbf{h}_{RS}^\dagger \mathbf{v} \qquad (4.24)$$

$$s.t. \ \mathbf{v}^\dagger \mathbf{v} \le P_{max} \qquad (4.25)$$

To achieve the maximum transmission rate, $\mathbf{v}$ should lie in the space spanned by $\mathbf{h}_{RS}$. Thus, $\mathbf{v}^\star$ can be given by $\mathbf{v}^\star = \sqrt{P_{max}} \frac{\mathbf{h}_{RS}}{\|\mathbf{h}_{RS}\|}$, where $\| \mathbf{h}_{RS} \|$ is the Euclidean norm of $\mathbf{h}_{RS}$. Therefore, given a certain α, the overall transmission rate $\bar{R}_S$ is given by

$$\bar{R}_S = (1 - \alpha) W \log_2 \left(1 + \frac{P_{max} \| \mathbf{h}_{RS} \|^2}{N_0 W}\right). \qquad (4.26)$$

Secrecy Rate Maximization

Similar to Sects. 4.3.1.2 and 4.3.1.3, the cluster of SUs, as a whole, determines the maximum power P_{max}^C for cooperation based on the satisfaction obtained. Substituting (4.16) and (4.22) into (4.11), we can obtain $\bar{R}_{SEC}$. To maximize $\bar{R}_{SEC}$, the PU selects the optimal time allocation coefficients and the SUs determine the best beamforming weights under a total power constraint.

4.4.1.2 Cooperation Parameters Determination

Optimal Weight Selection

The SUs select the optimal weight $\mathbf{w}^\star$ to maximize the secrecy rate $\bar{R}_{SEC}$. From (4.11), given α, maximizing $\bar{R}_{SEC}$ is equivalent to maximizing $\frac{R_R + R_D}{R_R + R_E}$. Substituting (4.22) into the ratio, the optimal weight can be determined by solving the following problem.

$$\max_{\mathbf{w}} \frac{\Psi_D + \mathbf{w}^\dagger \mathbf{h}_{RD}\mathbf{h}_{RD}^\dagger \mathbf{w}}{\Psi_E + \mathbf{w}^\dagger \mathbf{h}_{RE}\mathbf{h}_{RE}^\dagger \mathbf{w}}$$

$$s.t.\ \ \mathbf{w}^\dagger \mathbf{w} \le P_{max}^C$$

where $\Psi_D = (\gamma + \xi)N_0 W$ and $\Psi_E = (\gamma + \delta)N_0 W$. Let us rewrite $\mathbf{w} = \sqrt{P_{max}^C}\,\hat{\mathbf{w}}$, where $\hat{\mathbf{w}}^\dagger \hat{\mathbf{w}} = 1$. The above problem is then transformed into the following form:

$$\max_{\mathbf{w}} \frac{\Psi_D + p\hat{\mathbf{w}}^\dagger \mathbf{h}_{RD}\mathbf{h}_{RD}^\dagger \hat{\mathbf{w}}}{\Psi_E + p\hat{\mathbf{w}}^\dagger \mathbf{h}_{RE}\mathbf{h}_{RE}^\dagger \hat{\mathbf{w}}} \tag{4.27}$$

$$s.t.\ \ \hat{\mathbf{w}}^\dagger \hat{\mathbf{w}} = 1, p \le P_{max}^C \tag{4.28}$$

To guarantee $\bar{R}_{SEC}$ to be positive, it is necessary that the numerator be greater than the denominator. Due to this necessary condition, the derivative of the objective function in (4.28) with respect to p is positive and $\bar{R}_{SEC}$ is maximized when $p = P_{max}^C$. Thus, the above optimization problem can be further rewritten as

$$\max_{\hat{\mathbf{w}}} \frac{\hat{\mathbf{w}}^\dagger \mathbf{Q}_{RD} \hat{\mathbf{w}}}{\hat{\mathbf{w}}^\dagger \mathbf{Q}_{RE} \hat{\mathbf{w}}} \tag{4.29}$$

$$s.t.\ \ \hat{\mathbf{w}}^\dagger \hat{\mathbf{w}} = 1 \tag{4.30}$$

where

$$\mathbf{Q}_{RD} = \frac{\Psi_D}{P_{max}^C}\mathbf{I} + \mathbf{h}_{RD}\mathbf{h}_{RD}^\dagger$$

and

$$\mathbf{Q}_{RE} = \frac{\Psi_E}{P_{max}^C}\mathbf{I} + \mathbf{h}_{RE}\mathbf{h}_{RE}^{\dagger}.$$

The problem in (4.30) is a generalized eigenvector problem and the optimal $\hat{\mathbf{w}}^{\star}$ is selected as the uniform eigenvector of $\mathbf{Q}_{RD}\mathbf{Q}_{RE}^{-1}$ corresponding to its largest eigenvalue. Therefore, given α, the optimal $\mathbf{w}^{\star} = \sqrt{P_{max}^C}\hat{\mathbf{w}}^{\star}$.

Time Allocation

Similar to Sect. 4.3.2.2, $\beta^{\star}$ can be determined by substituting (4.22) into (4.14), when optimal $\mathbf{w}$ is selected. The optimal $\alpha^{\star}$ can be determined by solving the following problem, when the optimal weights and β are selected.

$$\alpha^{\star} = \arg\max \alpha\beta(R_D - R_E) \tag{4.31}$$

Note that β, R_D, and R_E are all functions of α $(0 < \alpha < 1)$.

4.4.2 C-B Scheme for Multiple Eavesdroppers (CBME)

4.4.2.1 Problem Formulation

For the case of multiple eavesdroppers, the cooperation can be performed in a two-phase way, as shown in Fig. 4.2b. The operation in the first phase is the same as that in the previous cases and the transmission rate R_R is given in (4.16).

In the second phase, instead of relaying the PU's data and transmitting its own data in different phases, the cluster transmits $\mathbf{x}$ which is the sum of the weighted version of the PU's information symbol $\tilde{s}$ and its information symbol z with $E\{|z|^2\} = 1$. Therefore, $\mathbf{x}$ can be represented by $\mathbf{x} = \mathbf{w}\tilde{s} + \mathbf{v}z$, where $\mathbf{w}$ and $\mathbf{v}$ are the column vectors of the weights of all SUs for transmitting the PU's symbol and SUs' symbol, respectively. Then, the received signals $y_{D,2}$ and $\mathbf{y}_{E,2}$ at D and eavesdroppers in the second phase can be written respectively as:

$$y_{D,2} = \mathbf{h}_{RD}^{\dagger}\mathbf{w}\tilde{s} + \mathbf{h}_{RD}^{\dagger}\mathbf{v}z + n_{RD} \tag{4.32}$$

$$\mathbf{y}_{E,2} = \mathbf{H}_{RE}^{\dagger}\mathbf{w}\tilde{s} + \mathbf{H}_{RE}^{\dagger}\mathbf{v}z + \mathbf{n}_{RE} \tag{4.33}$$

where $\mathbf{H}_{RE}$ is the matrix of channel coefficients between the SUs and eavesdroppers, and $\mathbf{n}_{RE}$ is the noise vector at eavesdroppers. To transmit the PU's data and its own data simultaneously, the cluster utilizes the approach based on zero-forcing beamforming, which is similar to the work in [100]. By doing so, the SUs' transmission will not interfere with the concurrent transmission of the PU, and vice versa. To this end, $\mathbf{v}$ should be in the null space of $\mathbf{h}_{RD}^{\dagger}$ such that $\mathbf{h}_{RD}^{\dagger}\mathbf{v} = 0$ and $\mathbf{w}$ should be in the null space of $\mathbf{h}_{RS}^{\dagger}$ such that $\mathbf{h}_{RS}^{\dagger}\mathbf{w} = 0$. Therefore, the overall transmission rate $\bar{R}_S$ at the secondary receiver is

$$\bar{R}_S = (1 - \alpha)W \log_2 \left(1 + \frac{|\, \mathbf{h}_{RS}^{\dagger}\mathbf{v}\, |^2}{N_0 W} \right). \tag{4.34}$$

Different from the pervious case, it is not necessary to enforce the PU to grant a reasonable period of time to SUs due to the following reasons: (1) relaying PU's data and transmitting SUs' data occupy the same period, and hence, the PU itself will not just allocate a quite short duration for the second phase, which affects the PU's performance as well; and (2) the cluster can achieve the expected transmission rate $\bar{R}_{EX}$ on its own, i.e., $\bar{R}_S = \bar{R}_{EX}$, by choosing $\mathbf{w}$ and $\mathbf{v}$. Denote by P_1 and P_2 the transmission power for relaying the PU's data $\tilde{s}$ and transmitting its own data z, respectively, where $P_1 = \mathbf{w}^{\dagger}\mathbf{w}$ and $P_2 = \mathbf{v}^{\dagger}\mathbf{v}$. Since the cluster has a total power budget P_{max}, it holds that $P_1 + P_2 \le P_{max}$. To maximize the secrecy rate of the PU and guarantee the expected transmission rate $\bar{R}_{EX}$ of the SUs, the cluster chooses the suitable $\mathbf{w}$ and $\mathbf{v}$ under the total power constraint, while the PU determines α.

4.4.2.2 Cooperation Parameters Determination

For convenience, let $\mathbf{w} = \sqrt{P_1}\hat{\mathbf{w}}$ and $\mathbf{v} = \sqrt{P_2}\hat{\mathbf{v}}$, respectively, where $\hat{\mathbf{w}}^{\dagger}\hat{\mathbf{w}} = 1$ and $\hat{\mathbf{v}}^{\dagger}\hat{\mathbf{v}} = 1$. To select the optimal $\mathbf{w}^{\star}$ and $\mathbf{v}^{\star}$, we perform the following two steps: (1) determine the optimal $\hat{\mathbf{w}}^{\star}$ and $\hat{\mathbf{v}}^{\star}$ given P_1 and P_2; (2) select P_1 and P_2, based on the results of the previous step.

Step 1

We first determine the optimal $\hat{\mathbf{w}}^{\star}$ and $\hat{\mathbf{v}}^{\star}$. For $\hat{\mathbf{v}}$, the objective is to maximize the transmission rate at the secondary receiver, under the constraint of no interference at D. Therefore, the optimal $\hat{\mathbf{v}}^{\star}$ can be determined by solving the following optimization problem.

$$\max_{\hat{\mathbf{v}}} \; |\, \mathbf{h}_{RS}^{\dagger}\hat{\mathbf{v}}\, |^2 \tag{4.35}$$

$$s.t. \; \mathbf{h}_{RD}^{\dagger}\hat{\mathbf{v}} = 0 \tag{4.36}$$

$$\hat{\mathbf{v}}^{\dagger}\hat{\mathbf{v}} = 1 \tag{4.37}$$

From (4.36), it can be seen that $\hat{\mathbf{v}}$ is orthogonal to $\mathbf{h}_{RD}$, which means $\hat{\mathbf{v}}$ belongs to the subspace of $\mathbf{h}_{RD}^{\perp}$, i.e., the null space of $\mathbf{h}_{RD}$. To maximize the objective function in (4.35), the optimal $\hat{\mathbf{v}}^{\star}$ should be selected in the direction of the orthogonal projection of $\mathbf{h}_{RS}$ onto $\mathbf{h}_{RD}^{\perp}$. Thus, $\hat{\mathbf{v}}^{\star}$ can be determined as follows:

$$\hat{\mathbf{v}}^{\star} = \frac{(\mathbf{I} - \hat{\mathbf{h}}_{RD}\hat{\mathbf{h}}_{RD}^{\dagger})\mathbf{h}_{RS}}{\parallel (\mathbf{I} - \hat{\mathbf{h}}_{RD}\hat{\mathbf{h}}_{RD}^{\dagger})\mathbf{h}_{RS} \parallel}, \tag{4.38}$$

where $\mathbf{I} - \hat{\mathbf{h}}_{RD}\hat{\mathbf{h}}_{RD}^{\dagger}$ is the orthogonal projection onto $\mathbf{h}_{RD}^{\perp}$ and $\hat{\mathbf{h}}_{RD}$ is the normalized vector of $\mathbf{h}_{RD}$.

For $\hat{\mathbf{w}}$, the objective is to maximize the secrecy rate of the PU. Due to the presence of multiple eavesdroppers, it is typically difficult to obtain the optimal $\hat{\mathbf{w}}^{\star}$. Instead, a suboptimal solution is devised as follows. The cluster selects $\hat{\mathbf{w}}$ to null out the PU's information at all eavesdroppers,[2] i.e., $\mathbf{H}_{RE}^{\dagger}\hat{\mathbf{w}} = 0$. By doing so, the transmission rate at all eavesdroppers are zero. Thus, maximizing the secrecy rate is equivalent to maximizing R_D, which is given by

$$R_D = W \log_2\left(1 + \xi + P_2 \frac{|\, \mathbf{h}_{RD}^{\dagger}\hat{\mathbf{w}} \,|^2}{N_0 W}\right), \tag{4.39}$$

where ξ is the same as that in (4.2).

As mentioned before, $\mathbf{w}$ should also be in the null space of $\mathbf{h}_{RS}^{\dagger}$. Thus, the optimal $\hat{\mathbf{w}}^{\star}$ can be selected such that $|\, \mathbf{h}_{RD}^{\dagger}\hat{\mathbf{w}} \,|$ is maximized under the constraint that $\mathbf{H}_{RE}^{\dagger}\hat{\mathbf{w}} = 0$ and $\mathbf{h}_{RS}^{\dagger}\mathbf{w} = 0$. Define a matrix $\mathbf{H}_R$, which contains $\mathbf{h}_{RS}$ and $\mathbf{H}_{RE}$, i.e., $\mathbf{H}_R = [\mathbf{h}_{RS}\ \mathbf{H}_{RE}]$. Then, the constraint becomes $\mathbf{H}_R^{\dagger}\mathbf{w} = 0$. To satisfy it, $\hat{\mathbf{w}}$ should belong to the subspace of $\mathbf{H}_R^{\perp}$, i.e., the null space of $\mathbf{H}_R$. To maximize $|\, \mathbf{h}_{RD}^{\dagger}\hat{\mathbf{w}} \,|$, the optimal $\hat{\mathbf{w}}^{\star}$ should be closest to $\mathbf{h}_{RD}^{\dagger}$ and belongs to $\mathbf{H}_R^{\perp}$. Thus, $\hat{\mathbf{w}}^{\star}$ should be the orthogonal projection of $\mathbf{h}_{RD}$ onto the subspace $\mathbf{H}_R^{\perp}$. Then, $\hat{\mathbf{w}}^{\star}$ can be derived as

$$\hat{\mathbf{w}}^{\star} = \frac{(\mathbf{I} - \mathbf{H}_R(\mathbf{H}_R^{\dagger}\mathbf{H}_R)^{-1}\mathbf{H}_R^{\dagger})\mathbf{h}_{RD}}{\parallel (\mathbf{I} - \mathbf{H}_R(\mathbf{H}_R^{\dagger}\mathbf{H}_R)^{-1}\mathbf{H}_R^{\dagger})\mathbf{h}_{RD} \parallel}, \tag{4.40}$$

where $\mathbf{H}_R(\mathbf{H}_R^{\dagger}\mathbf{H}_R)^{-1}\mathbf{H}_R^{\dagger}$ is the orthogonal projection matrix on $\mathbf{H}_R^{\perp}$.

Step 2

Determination of P_1, P_2 and α. Substituting (4.38) and (4.42) into (4.34) and (4.39), respectively, it can be seen that $\bar{R}_S$ is a function of P_1 and α, while R_D is a function of P_2. Given a certain α, the cluster needs to select P_1 to meet the expected transmission rate $\bar{R}_{EX}$ and the rest of power, i.e., P_2, contributes to R_D. Similar to

---

[2]Note that the number of SUs needs to be greater than that of eavesdroppers for this purpose.

the Appendix, when the secrecy rate is maximized, we have $\alpha = \frac{R_D}{R_R + R_D}$. Therefore, we have the following equations:

$$(1 - \alpha)W \log_2 \left(1 + \frac{P_1 \mid \mathbf{h}_{RS}^\dagger \hat{\mathbf{v}}^\star \mid^2}{N_0 W}\right) = \bar{R}_{EX} \qquad (4.41)$$

$$\alpha = \frac{R_D}{R_R + R_D} \quad P1 + P2 = P_{max}. \qquad (4.42)$$

Solving the above equations, we have

$$P_1 = \frac{(R_R N_0 + W\xi N_0 + \mid \mathbf{h}_{RD}^\dagger \hat{\mathbf{w}}^\star \mid^2)\bar{R}_{EX}}{R_R \mid \mathbf{h}_{RS}^\dagger \hat{\mathbf{v}}^\star \mid^2 + \mid \mathbf{h}_{RD}^\dagger \hat{\mathbf{w}}^\star \mid^2 \bar{R}_{EX}} \qquad (4.43)$$

$$\alpha = 1 - \frac{N_0 \bar{R}_{EX}}{P_1 \mid \mathbf{h}_{RS}^\dagger \hat{\mathbf{v}}^\star \mid^2} \qquad (4.44)$$

4.5   C-B Cooperation Scheme Without E-CSI (CBNE)

When E-CSI is unknown, it is impossible for the PU to determine the optimal length for the rewarding time, i.e., $(1 - \alpha)T$. Therefore, from the perspective of the PU, it desires that the SUs will make their best efforts to help for secure communication. To this end, the PU grants a time interval to SUs such that the need of SUs can be met, i.e., $\bar{R}_{EX}$ of the SUs can be obtained. In return, the SUs will make the best efforts to help the PU, i.e., to devote the maximum power P_{max} for cooperation.

4.5.1   Problem Formulation

The cooperation is carried out in a three-phase fashion, as shown in Fig. 4.2a. In the first phase, the transmission rate from S to the cluster and D are the same as in Sect. 4.4.1, which are given by (4.16) and (4.22), respectively.

In the second phase, all the cluster members transmit a combination of a weighted version of the re-encoded symbol $\tilde{s}$ and an artificial noise. Similar to [101], the artificial noise is leveraged to mask the concurrent transmission from S to D. As such, the cluster transmits $\mathbf{x}$, which is given by $\mathbf{x} = \mathbf{w}\tilde{s} + \mathbf{n}_a$, where $\mathbf{w}$ is the column vector of the weights of all SUs in the cluster and $\mathbf{n}_a$ is the artificial noise. Then, the received signals $y_{D,2}$ and $\mathbf{y}_{E,2}$ at D and eavesdroppers in the second phase can be written as:

$$y_{D,2} = \mathbf{h}_{RD}^\dagger \mathbf{w}\tilde{s} + \mathbf{h}_{RD}^\dagger \mathbf{n}_a + n_{RD} \qquad (4.45)$$

$$\mathbf{y}_{E,2} = \mathbf{h}_{RE}^\dagger \mathbf{w}\tilde{s} + \mathbf{h}_{RE}^\dagger \mathbf{n}_a + \mathbf{n}_{RE} \qquad (4.46)$$

As mentioned before, the total power constraint of the cluster for cooperation is P_{max}. Denote the power spent in transmitting the information symbol $\tilde{s}$ and the artificial noise $\mathbf{n}_a$ by P_I and P_N, respectively. It holds that $P_I + P_N \leq P_{max}$. To enhance the security of the PU, the cluster has to allocate the power properly.

Due to the unknown CSI related to the eavesdroppers, the cluster performs in the following way. In order to avoid interfering with D, the artificial noise should be transmitted in the null space of $\mathbf{h}_{RD}$ such that $\mathbf{h}_{RD}^{\dagger}\mathbf{n}_a = 0$. Moreover, instead of transmitting in certain dimension, the power of artificial noise should be spread uniformly in the dimensions of the null space of $\mathbf{h}_{RD}$ [98]. Since the artificial noise does not interfere with D but the eavesdroppers, more power allocated to the artificial noise is more beneficial to increasing the secrecy rate. However, allocating all the power to the artificial noise will cause that the transmission rate at D becomes extremely low, which is not desired. To avoid this, the power allocated to information symbol transmission, i.e., $\mathbf{w}^{\dagger}\mathbf{w}$, should guarantee that the transmission rate at D is above a predefined required transmission rate, which is similar to the work in [92]. Denote this predefined rate by $\bar{R}_Q$ and $\bar{R}_D$ should be greater than $\bar{R}_Q$ in order to meet this requirement. Therefore, the cluster allocates the minimum power for the information symbol transmission to achieve $\bar{R}_Q$ so that more power can be left to be utilized to confound the eavesdroppers.

The last phase is the same as that in Sect. 4.4.1 and the overall transmission rate $\bar{R}_S$ can be expressed as in (4.26), for a given α.

4.5.2 Cooperation Parameters Determination

4.5.2.1 Optimal Weight Selection

To achieve the above goal, we first determine the minimum power for $\bar{R}_Q$, which can be obtained by solving the following problem:

$$\min_{\mathbf{w}} \quad \mathbf{w}^{\dagger}\mathbf{w} \tag{4.47}$$

$$s.t. \quad \alpha W \log_2\left(1 + \xi + \frac{\mathbf{w}^{\dagger}\mathbf{h}_{RD}\mathbf{h}_{RD}^{\dagger}\mathbf{w}}{N_0 W}\right) \geq \bar{R}_Q, \tag{4.48}$$

where ξ is the same as in (4.2). The left hand side of the constraint is the overall transmission rate, which equals to α multiplied by R_D in (4.22). The inequality constraint yields the same result as the equality constraint. Thus, for the low SNR regime, the constraint can be further represented by

$$\mathbf{w}^{\dagger}\mathbf{h}_{RD}\mathbf{h}_{RD}^{\dagger}\mathbf{w} = \vartheta, \tag{4.49}$$

where $\vartheta = N_0 W(\frac{R_Q}{\alpha W} - \xi)$. Defining $\tilde{\mathbf{H}} = \mathbf{h}_{RD}\mathbf{h}_{RD}^{\dagger}$ and applying the method of Lagrange multipliers, the Lagrange multiplier function is given by

$$L(\mathbf{w}, \lambda) = \mathbf{w}^\dagger \mathbf{w} - \lambda(\mathbf{w}^\dagger \tilde{\mathbf{H}} \mathbf{w} - \vartheta), \tag{4.50}$$

where λ is the Lagrange multiplier. Take the derivative of $L(\mathbf{w}, \lambda)$ with respect to $\mathbf{w}^\dagger$, and let it be equal to zero. Then, we have $\tilde{\mathbf{H}}\mathbf{w} = \frac{\mathbf{w}}{\lambda}$. It can be seen that $1/\lambda$ is the eigenvalue of $\tilde{\mathbf{H}}$, while $\mathbf{w}$ is the corresponding eigenvector. Multiplying both sides of this equation by $\mathbf{w}^\dagger \lambda$, we can obtain

$$\mathbf{w}^\dagger \mathbf{w} = \lambda \mathbf{w}^\dagger \tilde{\mathbf{H}} \mathbf{w} = \lambda \vartheta, \tag{4.51}$$

where the last equality holds due to the constraint in (4.49). It can be seen that minimizing the transmission power, i.e., $\mathbf{w}^\dagger \mathbf{w}$, is equivalent to minimizing λ or to maximizing $1/\lambda$, since ϑ is a constant. Therefore, the optimal $\mathbf{w}^\star$ should be selected as the eigenvector of $\tilde{\mathbf{H}}$ corresponding to its largest eigenvalue. In other words, $\mathbf{w}^\star$ can be given by $\mathbf{w} = \varsigma \mathbf{n}$, where $\mathbf{n}$ is the normalized principal eigenvector of $\tilde{\mathbf{H}}$ and the scalar ς is given by $\varsigma = \sqrt{\frac{\vartheta}{\mathbf{n}^\dagger \tilde{\mathbf{H}} \mathbf{n}}}$. With $\mathbf{w}^\star$, the cluster spends the minimum power to meet the QoS requirement, and then, more power can be utilized to spread the artificial noise to confound the eavesdroppers.

4.5.2.2 Time Allocation

$\beta^\star$ can be determined by substituting (4.22) into (4.14), when the optimal $\mathbf{w}^\star$ is selected. The PU selects α such that the SUs can achieve the expected transmission rate and in return the SUs make their best efforts to help the PU. The overall transmission rate $\bar{R}_S$ at the secondary receiver is given in (4.26). To achieve $\bar{R}_{EX}$, α can be determined as

$$\alpha = 1 - \frac{\bar{R}_{EX}}{P_{max} \parallel \mathbf{h}_{RS} \parallel^2} \tag{4.52}$$

4.6 Simulation Results

In this section, we present simulation results to provide insight of the proposed cooperation schemes. In the simulation, bandwidth W is set to 1 Hz, and P_{max} and noise power are set to 2 and 1 mw, respectively. For R-J cooperation scheme, Fig. 4.3 shows the analytical (Ana.) and simulation (Sim.) results of the optimal transmission power of the jammer SU with respect to $|h_{JE}|^2$, when $|h_{RD}|^2$ is set to 0.6, 0.7 and 0.8, respectively, given that $|h_{JD}|^2 = 0.3$ and $|h_{RE}|^2 = 0.4$. It can be seen that the analytical and simulation results match each other well. It can also be seen that the optimal transmission power increases as $|h_{JE}|^2$ becomes greater and reduces when $|h_{RD}|^2$ increases. The reason is that when the channel between J and E is better, the

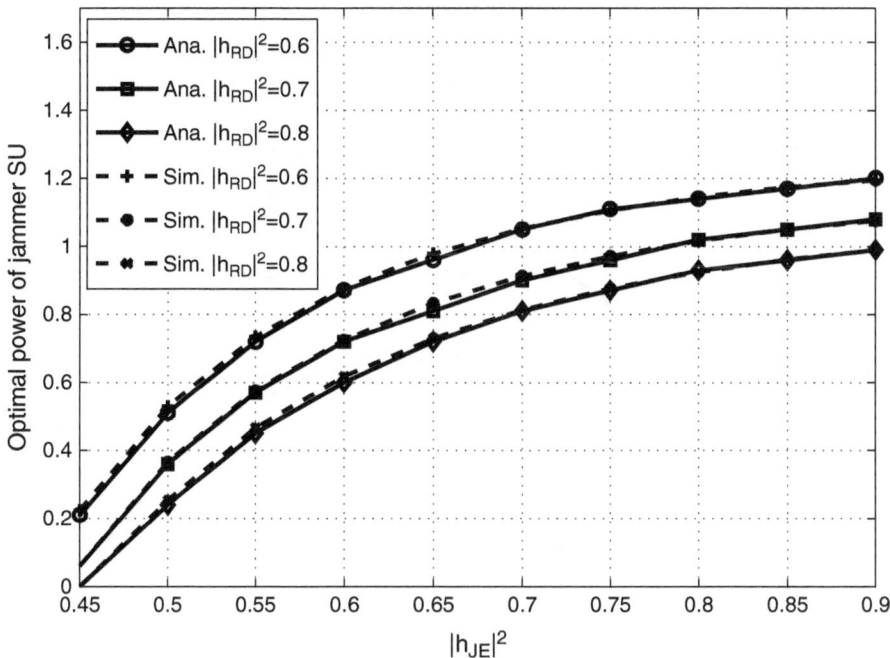

Fig. 4.3 Optimal transmit power of J for different h_{JE}

jammer SU is more beneficial and more transmission power should be allocated to the jamming signal to increase the secrecy rate.

Figure 4.4 shows the trends of the overall secrecy rate $\bar{R}_{SEC}$ of the PU with respect to the time allocation coefficient α, for different channel h_S between the SU and its corresponding receiver. In this simulation, $|h_{RD}|^2$, $|h_{RE}|^2$, $|h_{JD}|^2$, and $|h_{JE}|^2$ are set to 0.8, 0.5, 0.4, and 0.8, respectively. $|h_S^2|$ is chosen as 0.4, 0.6, and 0.8, respectively, while the expected transmission rate of SUs is 0.4 bit/s/Hz. It can be seen that $\bar{R}_{SEC}$ first increases and then decreases with α increasing. Furthermore, the maximum $\bar{R}_{SEC}$ is circled for the three lines and the corresponding optimal $\alpha^\star$ is 0.5, 0.55, and 0.6, respectively. In addition, both $\bar{R}_{SEC}$ and the optimal $\alpha^\star$ increase when the channel gain $|h_S|$ increases. This is because a better channel condition between the SU and its corresponding receiver will results in a better transmission rate, and hence, the PU can allocate a shorter period of time to SUs to achieve the same level of SUs' efforts, or the SUs are willing to devote more transmission power for cooperation given the same rewarding time.

Figure 4.5 shows $\bar{R}_{SEC}$ of the PU obtained by using R-J cooperation scheme and equal-duration cooperation scheme (EDRJ). The only difference between R-J scheme and EDRJ is that the time durations for the first two phases in EDRJ are equal and the secrecy rate is maximized without considering time allocation. The parameters $|h_{RE}|^2$, $|h_{JD}|^2$, $|h_{SD}|^2$, $|h_{SR}|^2$, $|h_{SE}|^2$ are set to 0.3, 0.3, 0.6, 0.3,

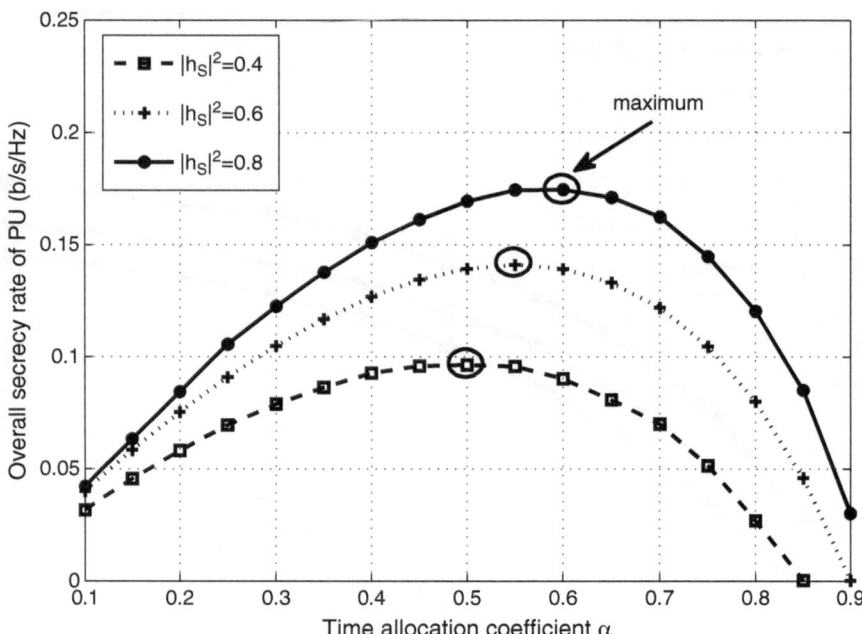

Fig. 4.4 Overall secrecy rate of PU versus α for R-J scheme

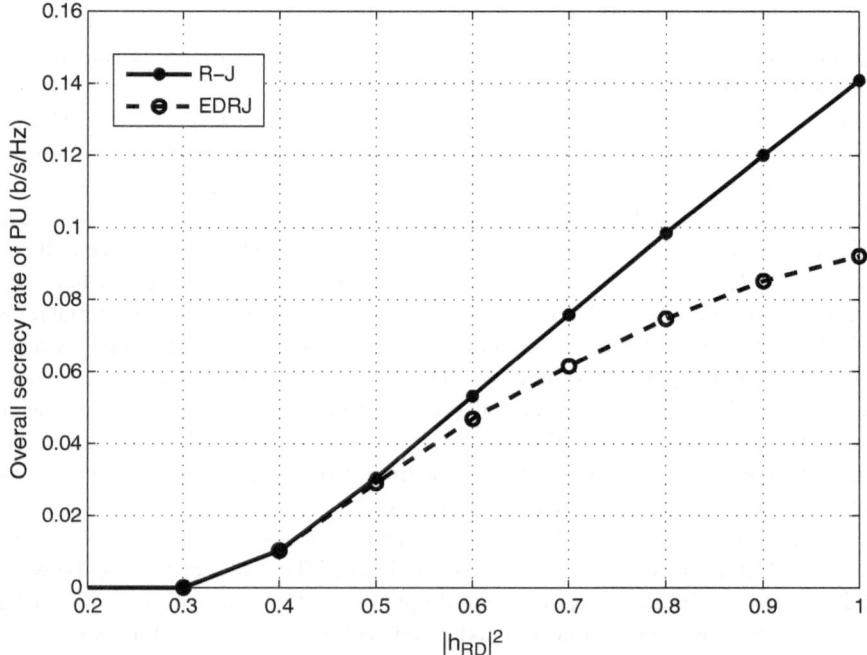

Fig. 4.5 Comparison between R-J scheme and EDRJ

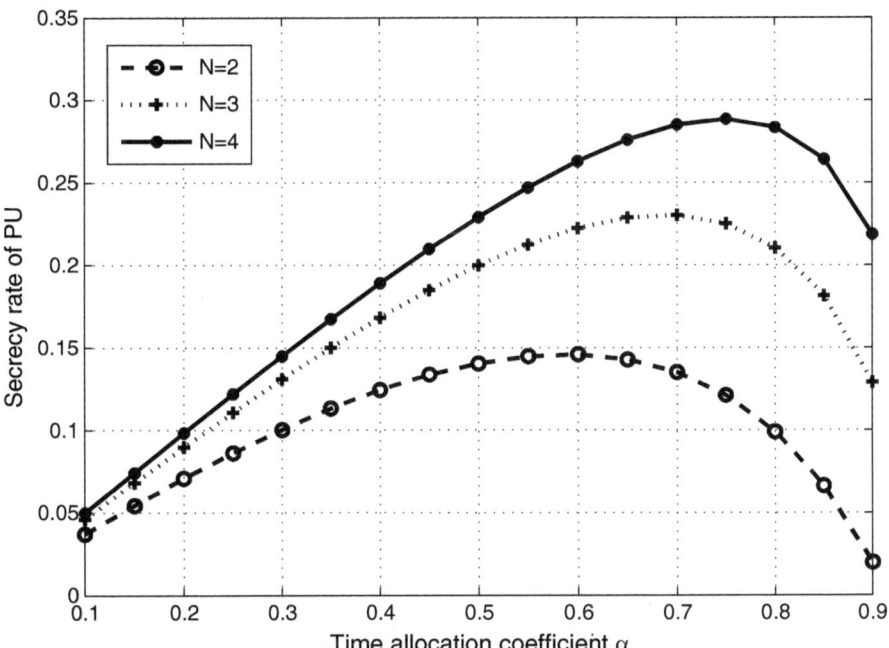

Fig. 4.6 Overall secrecy rate of PU versus α (CBSE)

0.4, respectively. It can be seen that R-J cooperation scheme outperforms EDRJ because R-J cooperation scheme jointly optimizes the time and transmission power to maximize $\bar{R}_{SEC}$. Moreover, the gap between them increases as h_{RD} increases.

Figure 4.6 shows $\bar{R}_{SEC}$ of the PU when cooperating with a cluster of SUs. $|h_{SD}|^2$ and $|h_{SE}|^2$ are set to 0.3 and 0.4, respectively, while the worst channel $|h_{SR,i}|^2$ is set to 0.4. For simplicity, the complex channels between all the SUs and D are approximately the same and equal to $e^{j\frac{\pi}{4}}$; similarly the complex channels between all the SUs and E are set to $0.8e^{j\frac{\pi}{4}}$. It can be seen that there exists an optimal $\alpha^\star$ such that $\bar{R}_{SEC}$ can achieve the maximum value. Moreover, $\bar{R}_{SEC}$ increases when the total number of SUs (N) in the cluster increases.

To evaluate the performance of C-B cooperation scheme, the complex channel coefficient h is given by $|h| \cdot e^{j\theta}$, where $|h|$ is the channel gain and θ is uniformly distributed in $[0, 2\pi)$. We obtain the average results using Monte Carlo simulation which consists of 1,000 trials. Figure 4.7 shows the maximum overall secrecy rate of the PU with respect to $|h_{RD}|$. The maximum $\bar{R}_{SEC}$ is calculated when the optimal parameters are selected. It can be seen that the maximum $\bar{R}_{SEC}$ increases when $|h_{RD}|$ increases and a smaller $|h_{RE}|$ leads to a larger $\bar{R}_{SEC}$.

Figure 4.8 shows $\bar{R}_{SEC}$ of the PU obtained by using C-B cooperation scheme and equal-duration cooperation scheme (EDCS) in the presence of an eavesdropper. The only difference between EDCS and CBSE is that the time durations for the first two

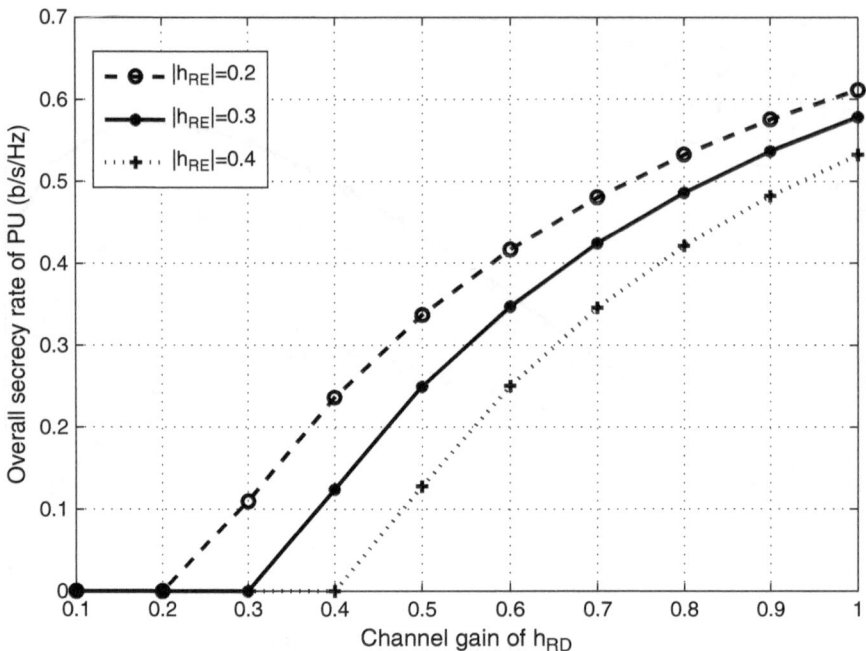

Fig. 4.7 Maximum overall secrecy rate of PU versus $|h_{RD}|$ (CBSE)

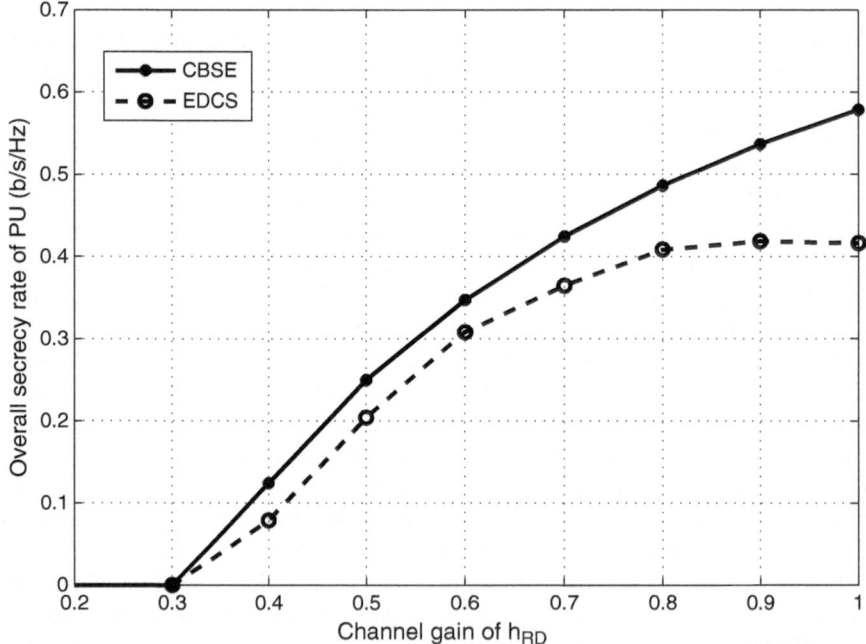

Fig. 4.8 Comparison between CBSE and EDCS

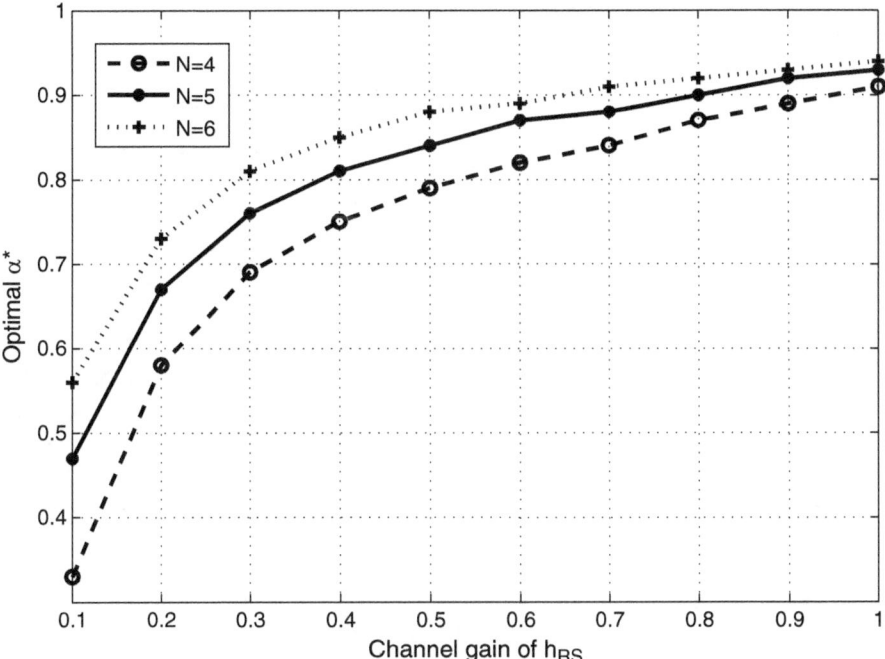

Fig. 4.9 Optimal α^* versus $|h_{RD}|$ (CBSE)

phases are equal in EDCS and the secrecy rate is maximized without considering time allocation. The parameters $|h_{RE}|$, $|h_{SD}|$, $|h_{SE}|$, N, are set to 0.3, 0.3, 0.4, 3, respectively, while the worst channel $|h_{SR,i}|^2$ is set to 0.4. It can be seen that C-B cooperation scheme outperforms EDCS because C-B cooperation scheme jointly optimizes the time and beamforming weights to maximize $\bar{R}_{SEC}$. Moreover, the gap between them increases as h_{RD} increases.

Figure 4.9 shows the trend of the optimal α^* with respect to $|h_{RS}|$ for different number of SUs in the cluster. In this simulation, $|h_{RE}|$ is set to 0.3, while $|h_{RD}|$ is selected as 0.4 such that the secrecy rate is positive. In addition, $\bar{R}_{EX}$ is set to 0.7 bit/s/Hz. It can be seen that the optimal α^* rises as $|h_{RS}|$ increases. The reason is similar to that for Fig. 4.4.

Figure 4.10 shows the overall secrecy rate of the PU with respect to the number of eavesdroppers (M) for different expected transmission rate of SUs. For this case, $|h_{RE}|$, $|h_{RD}|$, and $|h_{RS}|$ are set to 0.4, 0.5, and 0.6, respectively. In addition, the number of SUs (N) is set to 10. It can be seen that $\bar{R}_{SEC}$ drops as M increases. Moreover, it can also be seen that a lower $\bar{R}_{EX}$ results in a larger overall secrecy rate. This is because the SUs can spend less transmission power to achieve a lower $\bar{R}_{EX}$ and hence more power can be used to increase the secrecy rate of the PU.

Figure 4.11 shows the minimum transmission power of SUs with respect to $|h_{RD}|$ for different expected transmission rate of SUs. In this simulation,

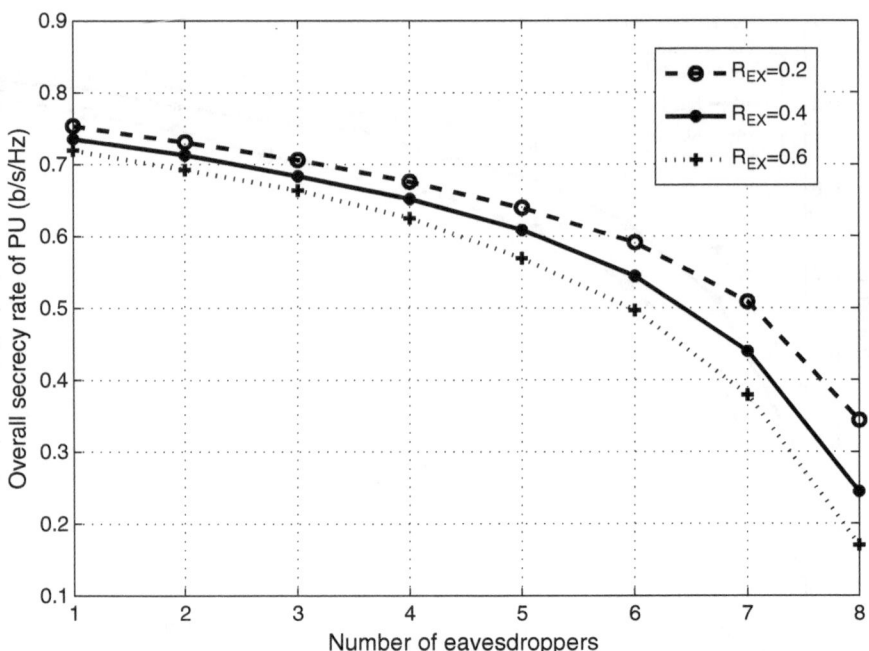

Fig. 4.10 $\bar{R}_{SEC}$ of PU versus the number of eavesdroppers (CBME)

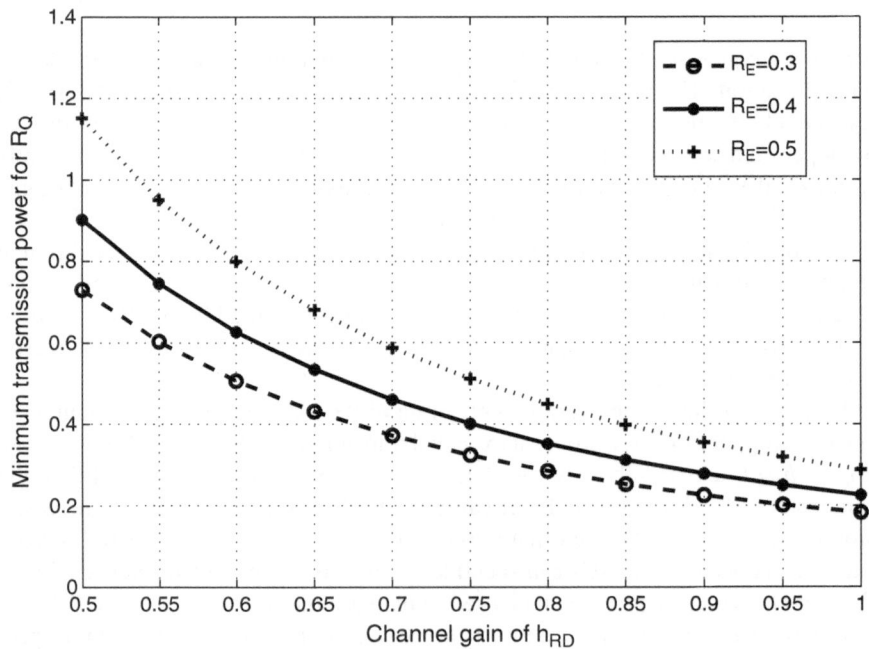

Fig. 4.11 Minimum transmission power versus $|h_{RD}|$ (CBNE)

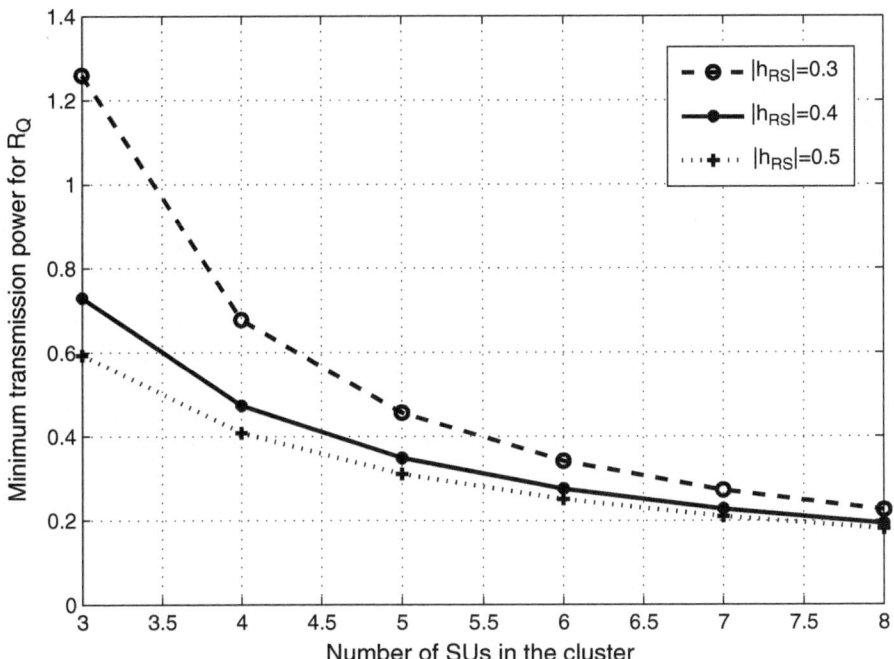

Fig. 4.12 Minimum transmission power versus the number of SUs (CBNE)

$|h_{RE}|$ is set to 0.4, while $|h_{RD}|$ is selected from 0.5 to 1. In addition, the required transmission rate $\bar{R}_Q = 0.5$ bit/s/Hz and $N = 3$. It can be seen that the minimum transmission power drops as $|h_{RD}|$ increases. This is because SUs can spend less transmission power to achieve the same QoS requirement, with a better channel condition. It can also be seen that a smaller $\bar{R}_{EX}$ results in a lower transmission power. The reason is that only a shorter time is needed for SUs to achieve a smaller $\bar{R}_{EX}$, which causes a larger α; and then, the SUs can spend less transmission power to achieve the same $\bar{R}_Q$.

Figure 4.12 shows the trends of the minimum transmission power of SUs versus the number of SUs in the cluster. $|h_{RD}|$ is set to 0.5 while $\bar{R}_{EX}$ is selected as 0.3 bit/s/Hz. It can be seen that the minimum transmission power drops as the number of SUs increases. Moreover, a smaller $|h_{RS}|$ results in a larger transmission power. That is because a longer duration for rewarding time is needed for SUs to achieve $\bar{R}_{EX}$ when $|h_{RS}|$ is smaller, which causes a smaller α; hence, the SUs need to spend more transmission power to help the PU to satisfy the QoS requirement.

4.7 Summary

In this chapter, we have proposed two types of cooperative spectrum access to enhance the security of the PU and provide channel access opportunities to SUs. In order to enhance the security, the PU can either cooperate with two individual SUs (R-J scheme) or a cluster of SUs (C-B scheme). For R-J scheme, the two SUs act as one relay and one friendly jammer to increase the secrecy rate of the PU in the presence of one eavesdropper. For C-B scheme, a cluster of SUs enhance the secrecy of the PU's communication via collaborative beamforming. Especially, for C-B scheme, three cooperation approaches have been proposed for the scenarios with one eavesdropper, with multiple eavesdroppers, and without any information about eavesdroppers. To maximize the secrecy rate, joint time and transmission power allocation is considered in R-J scheme, while time allocation and weight selection are jointly optimized in C-B schemes. We have shown through simulation results that with the proposed schemes, the secrecy of PU's communications can be significantly enhanced and the SUs can acquire certain access time.

Chapter 5
Concluding Remarks

In this book, we have investigated the security aspects of cooperation in cognitive radio networks. In particular, we have

- studied the basic concepts of cognitive radio, including the functions, network architecture, applications, access modes, and so on. The sensing techniques are summarized and the limitations of sensing are highlighted. Two forms of cooperation in CRNs are introduced. A survey on cooperative networking in CRNs is provided and the security aspects are discussed.
- proposed a novel cooperation scheme in CRNs to improve PUs' throughput or energy efficiency, considering the trustworthiness of SUs. The PU chooses the best cooperating SU and optimal cooperation parameters to maximize its utility based on the channel quality. While the selected SU determines the optimal power for cooperative and secondary transmissions. We formulate the procedure of decision making as a Stackelberg game. The analysis of the game provides the PU with the best strategy for partner selection, spectrum access time allocation and transmission power determination. Numerical results have demonstrated that, with the proposed scheme, the PU can obtain high throughput or energy saving through the cooperation by selecting the most suitable SU.
- investigated cooperation in CRNs, taking the physical layer security into consideration. With the proposed cooperation schemes, the PU enhances the security of communications and SUs can gain transmission opportunities. Particularly, the PU can either cooperate with two individual SUs or a cluster of SUs. For the former (R-J cooperation scheme), the two SUs act as one relay and one friendly jammer to increase the secrecy rate of the PU in the presence of one eavesdropper. For the latter (C-B cooperation scheme), a cluster of SUs enhances the secrecy of the PU's communication via collaborative beamforming. Three different scenarios with one eavesdropper, with multiple eavesdroppers, and without any information about eavesdroppers, are studied, respectively. To maximize the secrecy rate, joint time and transmit power allocation is considered in R-J cooperation scheme, while time allocation and weight selection are jointly

N. Zhang and J.W. Mark, *Security-aware Cooperation in Cognitive Radio Networks*, SpringerBriefs in Computer Science, DOI 10.1007/978-1-4939-0413-6_5, © The Author(s) 2014

optimized in C-B cooperation schemes. Numerical results have demonstrated that, with the proposed schemes, the secrecy of PU's communications can be significantly enhanced through cooperation with SUs.

References

1. I. Akyildiz, W. Lee, M. Vuran, and S. Mohanty, "Next generation/dynamic spectrum access/cognitive radio wireless networks: a survey," *Computer Networks*, vol. 50, no. 13, pp. 2127–2159, 2006.
2. J. Mitola III and G. Maguire Jr, "Cognitive radio: making software radios more personal," *IEEE Personal Communications*, vol. 6, no. 4, pp. 13–18, 1999.
3. S. Haykin, "Cognitive radio: brain-empowered wireless communications," *IEEE Journal on Selected Areas in Communications*, vol. 23, no. 2, pp. 201–220, 2005.
4. F. C. Commission *et al.*, "Notice of proposed rule making and order: Facilitating opportunities for flexible, efficient, and reliable spectrum use employing cognitive radio technologies," *ET docket*, no. 03-108, p. 73, 2005.
5. I. Akyildiz, W. Lee, M. Vuran, and S. Mohanty, "A survey on spectrum management in cognitive radio networks," *IEEE Communications Magazine*, vol. 46, no. 4, pp. 40–48, 2008.
6. Y.-C. Liang, Y. Zeng, E. C. Peh, and A. T. Hoang, "Sensing-throughput tradeoff for cognitive radio networks," *IEEE Transactions on Wireless Communications*, vol. 7, no. 4, pp. 1326–1337, 2008.
7. A. Ghasemi and E. Sousa, "Spectrum sensing in cognitive radio networks: requirements, challenges and design trade-offs," *IEEE Communications Magazine*, vol. 46, no. 4, pp. 32–39, 2008.
8. J. Huang, R. A. Berry, and M. L. Honig, "Auction-based spectrum sharing," *Mobile Networks and Applications*, vol. 11, no. 3, pp. 405–418, 2006.
9. H. Zhou, B. Liu, L. Gui, X. Wang, and Y. Li, "Fast spectrum sharing for cognitive radio networks: A joint time-spectrum perspective," in *Proc. of IEEE GLOBECOM*, 2011.
10. T. Han, T. Xing, N. Zhang, K. Liu, B. Tang, and Y. Liu, "Wireless spectrum sharing via waiting-line auction," in *Proc. of 11th IEEE ICCS*, 2008.
11. Q. Zhao and B. M. Sadler, "A survey of dynamic spectrum access," *IEEE Signal Processing Magazine*, vol. 24, no. 3, pp. 79–89, 2007.
12. Y. Xing, R. Chandramouli, S. Mangold *et al.*, "Dynamic spectrum access in open spectrum wireless networks," *IEEE Journal on Selected Areas in Communications*, vol. 24, no. 3, pp. 626–637, 2006.
13. Y. Wang, Y. Zhang, Q. Zhang, and S. Wu, "Optimal selection of false alarm probability for dynamic spectrum access," *IEEE Communications Letters*, vol. 17, pp. 844–847, 2013.
14. N. Cheng, N. Zhang, N. Lu, X. Shen, and J. W. Mark, "Opportunistic spectrum access for cr-vanets: A game theoretic approach," *IEEE Transactions on Vehicular Technology*, to appear.
15. Y. Wang, Q. Zhang, Y. Zhang, and P. Chen, "Adaptive resource allocation for cognitive radio networks with multiple primary networks," *EURASIP Journal on Wireless Communications and Networking*, vol. 2012, no. 1, pp. 1–18, 2012.

N. Zhang and J.W. Mark, *Security-aware Cooperation in Cognitive Radio Networks*, SpringerBriefs in Computer Science, DOI 10.1007/978-1-4939-0413-6,

16. J. Wang, M. Ghosh, and K. Challapali, "Emerging cognitive radio applications: A survey," *IEEE Communications Magazine*, vol. 49, no. 3, pp. 74–81, 2011.
17. D. Scaperoth, B. Le, T. Rondeau, D. Maldonado, C. W. Bostian, and S. Harrison, "Cognitive radio platform development for interoperability," in *Proc. of IEEE MILCOM*. IEEE, 2006, pp. 1–6.
18. I. Akyildiz, W. Lee, and K. Chowdhury, "Crahns: Cognitive radio ad hoc networks," *Ad Hoc Networks*, vol. 7, no. 5, pp. 810–836, 2009.
19. R. Manna, R. H. Louie, Y. Li, and B. Vucetic, "Cooperative spectrum sharing in cognitive radio networks with multiple antennas," *IEEE Transactions on Signal Processing*, vol. 59, no. 11, pp. 5509–5522, 2011.
20. Y. Zeng, Y. Liang, A. Hoang, and R. Zhang, "A review on spectrum sensing for cognitive radio: challenges and solutions," *EURASIP Journal on Advances in Signal Processing*, vol. 2010, 2010.
21. T. Yucek and H. Arslan, "A survey of spectrum sensing algorithms for cognitive radio applications," *IEEE Communications Surveys & Tutorials*, vol. 11, no. 1, pp. 116–130, 2009.
22. D. Cabric, S. Mishra, and R. Brodersen, "Implementation issues in spectrum sensing for cognitive radios," in *Proceedings of the 38th. Asilomar Conference on Signals, Systems, and Computers*, pp. 772–776, 2004.
23. H. Urkowitz, "Energy detection of unknown deterministic signals," *Proceedings of the IEEE*, vol. 55, no. 4, pp. 523–531, 1967.
24. F. Digham, M. Alouini, and M. Simon, "On the energy detection of unknown signals over fading channels," *IEEE Transactions on Communications*, vol. 55, no. 1, pp. 21–24, 2007.
25. W. Zhang, R. Mallik, and K. Letaief, "Optimization of cooperative spectrum sensing with energy detection in cognitive radio networks," *IEEE Transactions on Wireless Communications*, vol. 8, no. 12, pp. 5761–5766, 2009.
26. H. Kim and K. G. Shin, "In-band spectrum sensing in cognitive radio networks: energy detection or feature detection?" in *Proc. of ACM Mobicom*. ACM, 2008, pp. 14–25.
27. Y. Zeng, Y. C. Liang, and R. Zhang, "Blindly combined energy detection for spectrum sensing in cognitive radio," *IEEE Signal Processing Letters*, vol. 15, pp. 649–652, 2008.
28. Z. Ye, G. Memik, and J. Grosspietsch, "Energy detection using estimated noise variance for spectrum sensing in cognitive radio networks," in *Proc. of IEEE WCNC*. IEEE, 2008, pp. 711–716.
29. K.-L. Du and W. H. Mow, "Affordable cyclostationarity-based spectrum sensing for cognitive radio with smart antennas," *IEEE Transactions on Vehicular Technology*, vol. 59, no. 4, pp. 1877–1886, 2010.
30. Z. Ye, J. Grosspietsch, and G. Memik, "Spectrum sensing using cyclostationary spectrum density for cognitive radios," in *2007 IEEE Workshop on Signal Processing Systems*. IEEE, 2007, pp. 1–6.
31. I. Akyildiz, B. Lo, and R. Balakrishnan, "Cooperative spectrum sensing in cognitive radio networks: A survey," *Physical Communication*, 2010.
32. A. Ghasemi and E. Sousa, "Collaborative spectrum sensing for opportunistic access in fading environments," in *New Frontiers in Dynamic Spectrum Access Networks, 2005. DySPAN 2005. 2005 First IEEE International Symposium on*. IEEE, 2005, pp. 131–136.
33. W. Lee and I. Akyildiz, "Optimal spectrum sensing framework for cognitive radio networks," *IEEE Transactions on Wireless Communications*, vol. 7, no. 10, pp. 3845–3857, 2008.
34. N. Zhang, N. Cheng, H. Liang, Y. Tang, J. W. Mark, and X. Shen, "Efficient channel assignment for cooperative sensing based on convex bipartite matching," *Proceedings of IEEE ICC*, 2014.
35. J. Zhang and Q. Zhang, "Stackelberg game for utility-based cooperative cognitiveradio networks," in *Proceedings of the tenth ACM international symposium on Mobile ad hoc networking and computing*, 2009.
36. O. Simeone, I. Stanojev, S. Savazzi, Y. Bar-Ness, U. Spagnolini, and R. Pickholtz, "Spectrum leasing to cooperating secondary ad hoc networks," *IEEE Journal on Selected Areas in Communications*, vol. 26, no. 1, pp. 203–213, 2008.

37. N. Zhang, N. Cheng, N. Lu, H. Zhou, J. W. Mark, and X. Shen, "Cooperative cognitive radio networking for opportunistic channel access," in *Proceedings of IEEE GLOBECOM*, 2013.
38. E. Peh and Y.-C. Liang, "Optimization for cooperative sensing in cognitive radio networks," in *Proceedings of IEEE WCNC 2007*. IEEE, pp. 27–32.
39. G. Ganesan and Y. Li, "Cooperative spectrum sensing in cognitive radio networks," in *Proceedings of IEEE DySPAN*. IEEE, 2005, pp. 137–143.
40. S. Mishra, A. Sahai, and R. Brodersen, "Cooperative sensing among cognitive radios," in *Proc. of IEEE ICC*, vol. 4. IEEE, 2006, pp. 1658–1663.
41. Y. Zou, Y. Yao, and B. Zheng, "A selective-relay based cooperative spectrum sensing scheme without dedicated reporting channels in cognitive radio networks," *IEEE Transactions on Wireless Communications*, vol. 10, no. 4, pp. 1188–1198, 2011.
42. G. Ganesan and Y. Li, "Cooperative spectrum sensing in cognitive radio, part i: Two user networks," *IEEE Transactions on Wireless Communications*, vol. 6, no. 6, pp. 2204–2213, 2007.
43. Y. Zou, Y. Yao, and B. Zheng, "Cooperative relay techniques for cognitive radio systems: Spectrum sensing and secondary user transmissions," *IEEE Communications Magazine*, 2012.
44. R. Chen, J. Park, and J. Reed, "Defense against primary user emulation attacks in cognitive radio networks," *IEEE Journal on Selected Areas in Communications*, vol. 26, no. 1, pp. 25–37, 2008.
45. R. Chen, J. Park, and K. Bian, "Robust distributed spectrum sensing in cognitive radio networks," pp. 1876–1884, 2008.
46. P. Kaligineedi, M. Khabbazian, and V. Bhargava, "Secure cooperative sensing techniques for cognitive radio systems," pp. 3406–3410, 2008.
47. L. Lazos, S. Liu, and M. Krunz, "Mitigating control-channel jamming attacks in multi-channel ad hoc networks," pp. 169–180, 2009.
48. N. Zhang, N. Lu, N. Cheng, J. W. Mark, and X. Shen, "Cooperative spectrum access towards secure information transfer for crns," *IEEE Journal on Selected Areas in Communications*, to appear.
49. T. Elkourdi and O. Simeone, "Spectrum leasing via cooperation with multiple primary users," *IEEE Transactions on Vehicular Technology*, vol. 61, no. 2, pp. 820–825, 2012.
50. S. Hua, H. Liu, M. Wu, and S. Panwar, "Exploiting mimo antennas in cooperative cognitive radio networks," in *Proceedings IEEE INFOCOM*, Shanghai, China, April, 2011.
51. Y. Han, A. Pandharipande, and S. Ting, "Cooperative decode-and-forward relaying for secondary spectrum access," *IEEE Transactions on Wireless Communications*, vol. 8, no. 10, pp. 4945–4950, 2009.
52. Y. Yi, J. Zhang, Q. Zhang, T. Jiang, and J. Zhang, "Cooperative communication-aware spectrum leasing in cognitive radio networks," in *Proceedings of IEEE DySPAN 2010*. IEEE, pp. 1–11.
53. N. Zhang, N. Cheng, N. Lu, H. Zhou, J. W. Mark, and X. Shen, "Risk-aware cooperative spectrum access for multi-channel cognitive radio networks," in *IEEE Journal on Selected Areas in Communications*, 2013.
54. I. Stanojev, O. Simeone, U. Spagnolini, Y. Bar-Ness, and R. Pickholtz, "Cooperative arq via auction-based spectrum leasing," *IEEE Transactions on Communications*, vol. 58, no. 6, pp. 1843–1856, 2010.
55. B. Cao, L. Cai, H. Liang, J. W. Mark, Q. Zhang, H. Poor, and W. Zhuang, "Cooperative cognitive radio networking using quadrature signaling," in *Proceedings of IEEE INFOCOM*, Orlando, USA, March 2012.
56. B. Cao, Q. Zhang, J. W. Mark, L. X. Cai, and H. V. Poor, "Toward efficient radio spectrum utilization: User cooperation in cognitive radio networking," *IEEE Network*, vol. 26, no. 4, pp. 46–52, 2012.
57. N. Zhang, N. Lu, R. Lu, J. W. Mark, and X. Shen, "Energy-efficient and trust-aware cooperation in cognitive radio networks," *Proceedings of IEEE ICC*, 2012.

58. Y. Han, A. Pandharipande, and S. H. Ting, "Cooperative spectrum sharing via controlled amplify-and-forward relaying," pp. 1–5, 2008.
59. Y. Han, S. Ting, and A. Pandharipande, "Cooperative spectrum sharing protocol with secondary user selection," *IEEE Transactions on Wireless Communications*, vol. 9, no. 9, pp. 2914–2923, 2010.
60. Q. Li, S. H. Ting, A. Pandharipande, and Y. Han, "Cognitive spectrum sharing with two-way relaying systems," *IEEE Transactions on Vehicular Technology*, vol. 60, no. 3, pp. 1233–1240, 2011.
61. Y. Tang and J. W. Mark, "A quadrature signaling based cooperative scheme for cognitive radio networks," *Proceedings of IEEE Goblecom*, 2013.
62. N. Zhang, N. Lu, N. Cheng, J. W. Mark, and X. Shen, "Towards secure communications in cooperative cognitive radio networks," in *Proceedings of IEEE ICCC*, 2013.
63. Q. Li, S. Ting, A. Pandharipande, and M. Motani, "Cooperate-and-access spectrum sharing with arq-based primary systems," *IEEE Transactions on Communications*, vol. 60, Issue: 10, pp. 2861– 2871, 2012.
64. N. Michelusi, P. Popovski, O. Simeone, M. Levorato, and M. Zorzi, "Cognitive access policies under a primary arq process via forward-backward interference cancellation," *IEEE Journal on Selected Areas in Communications*, 2013.
65. J. C. Li, W. Zhang, A. Nosratinia, and J. Yuan, "Opportunistic spectrum sharing based on exploiting arq retransmission in cognitive radio networks," pp. 1–5, 2010.
66. D. Chiarotto, O. Simeone, and M. Zorzi, "Spectrum leasing via cooperative opportunistic routing techniques," *IEEE Transactions on Wireless Communications*, vol. 10, no. 9, pp. 2960–2970, 2011.
67. B. Cao, J. W. Mark, and Q. Zhang, "A polarization enabled cooperation framework for cognitive radio networking," *Proceedings of IEEE Goblecom*, 2012.
68. J.-H. Wui and D. Kim, "Cognitive relaying systems based on network and superposition coding in multiple access primary channels," pp. 1–5, 2011.
69. W. Su, J. D. Matyjas, and S. Batalama, "Active cooperation between primary users and cognitive radio users in cognitive ad-hoc networks," vol. 60, Issue: 4, pp. 1796–1805, 2012.
70. K. Lee, O. Simeone, C. Chae, and J. Kang, "Spectrum leasing via cooperation for enhanced physical-layer secrecy," in *Proc. of IEEE ICC'11*, 2011.
71. N. Zhang, N. Lu, N. Cheng, J. W. Mark, and X. Shen, "Cooperative networking towards secure communications for crns," in *Proc. of IEEE WCNC*, 2013.
72. Y. Wu and K. Liu, "An information secrecy game in cognitive radio networks," *IEEE Transactions on Information Forensics and Security*, vol. 6, no. 3, pp. 831–842, 2011.
73. D. Shila, Y. Cheng, and T. Anjali, "Mitigating selective forwarding attacks with a channel-aware approach in wmns," *IEEE Transactions on Wireless Communications*, vol. 9, no. 5, pp. 1661–1675, 2010.
74. H. V. Poor, "Information and inference in the wireless physical layer," *IEEE Wireless Communications*, vol. 19, no. 2, pp. 40–47, 2012.
75. L. Ozarow and A. Wyner, "Wire-tap channel ii," in *Advances in Cryptology*. Springer, pp. 33–50, 1985.
76. J. Li, A. Petropulu, and S. Weber, "On cooperative relaying schemes for wireless physical layer security," *IEEE Transactions on Signal Processing*, no. 99, pp. 1–1, 2011.
77. M. Gursoy, "Secure communication in the low-snr regime: A characterization of the energy-secrecy tradeoff," in *Proc. of IEEE ISIT'09*.
78. L. Dong, Z. Han, A. Petropulu, and H. V. Poor, "Improving wireless physical layer security via cooperating relays," *IEEE Transactions on Signal Processing*, vol. 58, no. 3, pp. 1875–1888, 2010.
79. M. Osborne, "An introduction to game theory," *Oxford University Press*, 2003.
80. A. Boukerche and Y. Ren, "A trust-based security system for ubiquitous and pervasive computing environments," *Computer Communications*, vol. 31, no. 18, pp. 4343–4351, 2008.

81. H. Yu, Z. Shen, C. Miao, C. Leung, and D. Niyato, "A survey of trust and reputation management systems in wireless communications," *Proceedings of the IEEE*, vol. 98, no. 10, pp. 1755–1772, 2010.

82. S. Ganeriwal, L. Balzano, and M. Srivastava, "Reputation-based framework for high integrity sensor networks," *ACM Transactions on Sensor Networks (TOSN)*, vol. 4, no. 3, pp. 1–37, 2008.

83. A. Jøsang and R. Ismail, "The beta reputation system," vol. 160, 2002.

84. Y. Mao and M. Wu, "Tracing malicious relays in cooperative wireless communications," *IEEE Transactions on Information Forensics and Security*, vol. 2, no. 2, pp. 198–212, 2007.

85. T. Khalaf and S. Kim, "Error probability in multi-source, multi-relay networks under falsified data injection attacks," pp. 1–4, 2008.

86. S. Dehnie, H. Senear, and N. Memon, "Detecting malicious behavior in cooperative diversity," in *41st Annual Conference on Information Sciences and Systems*. IEEE, 2007, pp. 895–899.

87. S. Dehnie and N. Memon, "Detection of misbehavior in cooperative diversity," in *Proceedings of IEEE MILCOM 2008*. IEEE, 2008, pp. 1–5.

88. R. Lu, X. Li, X. Liang, X. Shen, and X. Lin, "Grs: The green, reliability, and security of emerging machine to machine communications," *IEEE Communications Magazine*, vol. 49, pp. 28–35, 2011.

89. N. Anand, S. Lee, and E. Knightly, "Strobe: Actively securing wireless communications using zero-forcing beamforming," in *Proc. of IEEE INFOCOM'12*, 2012.

90. J. Huang and A. Swindlehurst, "Robust secure transmission in miso channels based on worst-case optimization," *IEEE Transactions on Signal Processing*, vol. 60, no. 4, pp. 1696–1707, 2012.

91. D. Goeckel, S. Vasudevan, D. Towsley, S. Adams, Z. Ding, and K. Leung, "Artificial noise generation from cooperative relays for everlasting secrecy in two-hop wireless networks," *IEEE Journal on Selected Areas in Communications*, vol. 29, no. 10, pp. 2067–2076, 2011.

92. H. Wang, Q. Yin, and X. G. Xia, "Distributed beamforming for physical-layer security of two-way relay networks," *IEEE Transactions on Signal Processing*, vol. 60, pp. 3532–3545, 2012.

93. L. Lai and H. El Gamal, "The relay–eavesdropper channel: Cooperation for secrecy," *IEEE Transactions on Information Theory*, vol. 54, no. 9, pp. 4005–4019, 2008.

94. G. Zheng, L. Choo, and K. Wong, "Optimal cooperative jamming to enhance physical layer security using relays," *IEEE Transactions on Signal Processing*, vol. 59, no. 3, pp. 1317–1322, 2011.

95. Z. Gao, Y. Yang, and K. Liu, "Anti-eavesdropping space-time network coding for cooperative communications," *IEEE Transactions on Wireless Communications*, vol. 10, no. 11, pp. 3898–3908, 2011.

96. L. Dong, Z. Han, A. Petropulu, and H. Poor, "Secure wireless communications via cooperation," in *Annual Allerton Conference on Communication, Control, and Computing*, 2008.

97. H. Ochiai, P. Mitran, H. Poor, and V. Tarokh, "Collaborative beamforming for distributed wireless ad hoc sensor networks," *IEEE Transactions on Signal Processing*, vol. 53, no. 11, pp. 4110–4124, 2005.

98. X. Gong, J. Wu, and J. Zhang, "Secure wireless communications via cooperative relaying and jamming," *Proceedings of IEEE GLOBECOM*, 2011.

99. G. Kim, *Scheduling in wireless ad hoc networks: algorithms with performance guarantees*. ProQuest, 2008.

100. A. Wiesel, Y. C. Eldar, and S. Shamai, "Zero-forcing precoding and generalized inverses," *IEEE Transactions on Signal Processing*, vol. 56, no. 9, pp. 4409–4418, 2008.

101. S. Goel and R. Negi, "Guaranteeing secrecy using artificial noise," *IEEE Transactions on Wireless Communications*, vol. 7, no. 6, pp. 2180–2189, 2008.